Valve Actuators

*A comprehensive introduction to
the design, selection, sizing, and application of
valve and damper actuators*

C. Warnett

CPLloyd
October 2015

VALVE ACTUATORS
A comprehensive introduction to
the design, selection, sizing, and application of
valve and damper actuators
C. Warnett

Copy Editor: Genilee Parente
Technical Editor: Marty Thomas
Book Designer: Kusmin
Illustrator: Palani

First Edition: October 2015

ISBN-13: 978-0692522035 (CPLloyd)
ISBN-10: 0692522034

Dedication

———— ◆ ————

This book is dedicated to the memory of Jeremy J Fry, who was an inspiration and mentor to many designers, engineers, artists and entrepreneurs.

———— ◆ ————

Contents

Foreword

Nobody knows more about valve automation than Chris Warnett. His involvement began after gaining a first class honors degree in engineering from the University of Bath, in the U.K.

Jeremy Fry, the founder of Rotork, the valve actuator manufacturer, had a keen eye for talented individuals, and he recruited Chris in the 1970s to work closely with him on new actuator designs.

Chris moved from engineering to sales, marketing and service support positions within that company. This span of roles gave him a unique perspective and insight into the full range of valve automation applications.

He has distilled 40 years of knowledge and the data he's put into many articles on the subject of actuation into this easy-to-use reference work on everything one needs to know about valve actuators.

The work covers the wide range of applications and environments that challenge those actuators. He covers the many and varied types of designs and has developed his SIMPLE guide to assist engineers and users in understanding what type of actuator is suitable for a particular valve and application.

This guide promises to be a valuable tool for anyone who needs to know about actuators.

Bill Whiteley
Chairman Spirax Sarco Engineering Plc and former CEO Rotork Plc.

Chapter 1

Introduction

— ◆ —

Most people are aware that we have become dependent on certain technologies in our daily lives, like computers, but are unaware that we are also dependent on automated valves. From the first waking moments of each day we draw on the basic utilities provided by an infrastructure of industrial plants containing automated valves. When we switch on a light, make coffee or drive to work, we are consuming products and services provided by electric power plants, potable water and waste water treatment plants, pulp mills, oil refineries, food processing plants, and many more.

All of these facilities use large quantities of valves to control their production processes. A large and growing number of these valves are automated.

A valve, in its most basic form, consists of a body and an internal moving component (closure element) that shuts off or restricts flow through the valve. To automate a valve, an actuator is added to the assembly to power the internal closure element of the valve.

A valve actuator then, is a mechanism for moving the closure element of a valve to a desired position. This could provide a basic isolation function, moving the valve to either the fully open or the fully closed position, or the valve actuator could position the valve closure element at any intermediate position to regulate the media flow through the pipeline.

Automated valves are employed under many differing circumstances, for example:

1) Where a number of valves have to be operated in sequence to achieve a coordinated movement of pipeline media.

2) For large valves that would otherwise take a long time, or many people to operate manually.

3) Where a valve is located in a hostile or hazardous location where manual operation would be difficult, restricted or otherwise undesirable.

4) In process control where repeated adjustments to the valve position are continuously required to achieve a desired production parameter.

5) In applications where the emergency shut-down of a valve or valves is necessary for environmental or personnel safety reasons.

This book is aimed at helping engineers and users understand the types of valves and actuators that are available, their characteristics and application. It will also help determine which type of actuator is suitable for a particular valve and its application. We will progress through the logical steps to make the right choice of valve actuator type and to eventually build a specification for that actuator. Further, if the specific valve size and details are known then the actuator size can also be selected.

I have named the method used to simplify the selection of the appropriate actuator for an application the "SIMPLE" method. This acronym helps us remember the key information required in the actuator selection process.

SIMPLE actuator selection:

SI for Size of the valve and its required force demand on the actuator plus the required speed of operation.

M is the Motion required of the actuator to move the valve's closure element and its frequency of movement (e.g., linear or rotary motion).

P is the Power source available, such as fluid power or electric power and the specifics of the power supply.

L is the Location where the actuator will be working and its constraints (e.g., space, temperature, etc.).

E is the Environment at that locale which will give us the enclosure requirements (e.g. hazardous, dust tight etc.).

Looking at the valve from the aspect of the actuator, there are far more variations of valve types than there are of actuators, especially when you consider the variations in design, trim, and body materials present in the valve manufacturing world. Fortunately, the valve actuator industry is able to automate the full spectrum of valves available with a comparatively smaller number of actuator variants.

Generally, the closure element of any valve is actuated either by a rotary or a linear motion. So regardless of the job that the valve is doing, the actuator will move the valve by one or other of these motions. This is obviously a simplification as we have not yet considered the magnitude of the force required, the source of power for the actuator, or many of the other parameters that impact actuator selection.

Any given actuator can automate many different types of valve, which has allowed a symbiotic automation industry to develop separately but parallel to the valve industry.

There are several major independent actuator manufacturing companies, while at the same time there are some actuator manufacturers that are part of valve maker companies.

In this book we will look at all the types of valve actuator, regardless of manufacture and type, and try to simplify the process of understanding and selecting the best actuator for a given application.

Chapter 2

Valves – from the actuators perspective

There are many types of valve in the industrial valve world. They can be categorized by type, industry, size, pressure class, or one of many other parameters.

To try to simplify the classification of valves, it may be easier to consider them from the viewpoint of the actuator. In order to automate a valve, the actuator really only needs to accommodate four major parameters: Force, Motion, Frequency and Speed.

When we select an actuator, we typically look at the force and motion requirements simultaneously, as in the SIMPLE method. However, when categorizing valves let's look at the motion first as this is defined by the design.

Examining each of these in turn:
1) Motion required by the actuator will take one of the following forms depending on the design of the valve and closure element:
    1) Multi-turn rotation
    2) Linear motion
    3) Part-turn rotation
2) The magnitude of the linear or rotary force required will be a function of the valve size, the mechanical design of the valve, and the physical

properties of the media such as pressure, temperature, and viscosity.

3) Speed of operation will be a function of the process requirements for the particular application.

4) Frequency of operation will depend on whether the valve is for an isolating, regulating, or modulating duty application. (We will examine these definitions later.)

Many valves have common components to which we will be referring throughout this book

1) Stem nut – the threaded nut that engages with a threaded valve stem.

2) Drive bush – the keyed or splined bushing that couples the actuator output to a rotating stem (multi or part-turn) of the valve.

3) Thrust base – the element of the valve that contains the thrust reaction from moving and seating the valve.

4) Top-works – the arrangement on top of the valve to which an actuator is attached.

5) Valve stem – the shaft protruding from the pressure containing body of the valve that is connected to the closure element.

6) Yoke – supports the top-works.

7) Gland packing – seals the valve stem.

8) Bonnet – top of the valve body, usually removable.

9) Seat – seals for the closure element.

10) Closure element – the internal moving part of a valve that blocks or restricts flow in the pipeline. For example, the wedge in a wedge gate valve, the sphere in a ball valve or the disc in a butterfly valve.

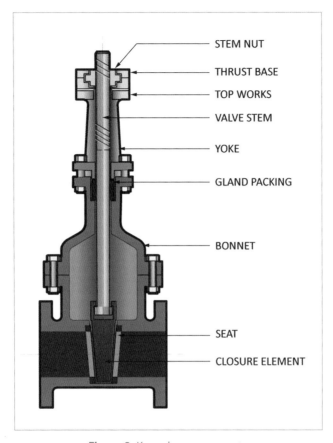

Figure 2. Key valve components.

2.1　Motion

Classifying valves by the first parameter, motion, we will soon see there are some distinct differences in valve designs.

2.1.1　Valves with linear motion of the closure element

Valves with a closure element requiring linear movement could use a fluid power linear cylinder or diaphragm coupled to a smooth stem. Alternatively, a multi-turn actuator with an electric or even fluid-powered motor could be used to drive a threaded stem.

To allow multi-turn actuators to automate them, valves with a linear motion on the closure element need a rotary to linear conversion mechanism. This conversion mechanism is described as one of the following categories of valve stem designations:

1) Rising non-rotating stem
2) Rotating non-rising stem
3) Rising rotating stem

2.1.1.1   The rising non-rotating stem valve

This is typically a gate valve that has a fixed threaded stem. A matching threaded nut in the actuator (stem nut), when rotated, will convert the rotary motion to a linear motion and will move the stem up or down. The stem is connected to the disc or plug (closure element) in the valve and the actuator can open or close the valve by rotating the stem nut in a clockwise or counterclockwise direction. As the stem nut is located in the actuator "thrust base" assembly, the thrust of the valve is also contained in the actuator.

Some actuator designs may have a detachable thrust base. This has the benefit of allowing the actuator to be removed more easily from the valve for service. The thrust base can remain on the valve, holding the stem and closure element in positon, allowing the valve to remain in static service.

Also, heavier actuators can be mounted more easily if the thrust base is a separate assembly.

The rising non-rotating stem type is a very common valve design for valves used in the power, water, and the oil and gas industries.

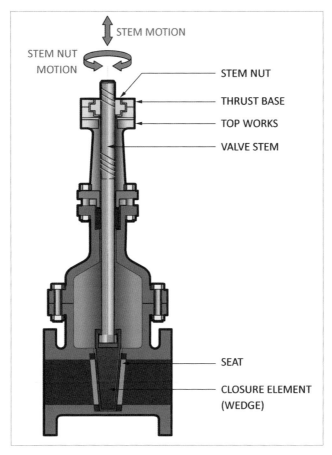

Figure 2.1.1.1. Rising non-rotating stem valve, gate valve.

2.1.1.2   The rotating non-rising stem valve

This valve works on a similar principle to the rising non-rotating stem valve, but has the stem nut located in the closure element of the valve. This is more complex as the valve design has to allow for the length of the threaded stem to fit inside the closure element as it travels to the open position. The valve thrust is contained in a thrust bearing in the valve top-works. This type of stem arrangement is often used on gate valves in the municipal water and waste industries.

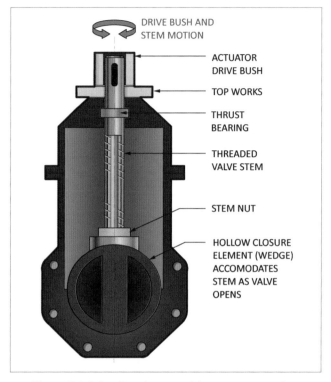

DRIVE BUSH AND
STEM MOTION

ACTUATOR
DRIVE BUSH

TOP WORKS

THRUST
BEARING

THREADED
VALVE STEM

STEM NUT

HOLLOW CLOSURE
ELEMENT (WEDGE)
ACCOMODATES
STEM AS VALVE
OPENS

Figure 2.1.1.2.a. Rotating non-rising stem, gate valve.

2.1.1.3   The rising rotating stem

This valve is a hybrid of the previous designs. The stem nut is located in the top of the valve so as the stem is rotated, it raises or lowers the closure element. The actuator is coupled to the valve stem with a sliding spline or "butterfly" nut arrangement. Typically, the thrust of the valve is contained by the top-works of the valve. This type of valve is popular with boiler makers for drain and steam lines.

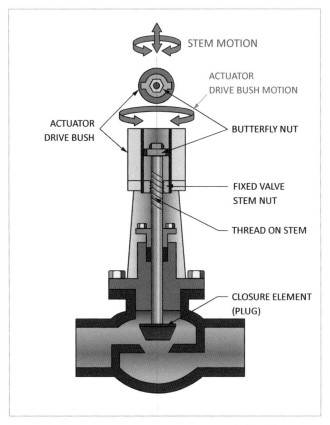

STEM MOTION

ACTUATOR
DRIVE BUSH MOTION

ACTUATOR
DRIVE BUSH

BUTTERFLY NUT

FIXED VALVE
STEM NUT

THREAD ON STEM

CLOSURE ELEMENT
(PLUG)

Figure 2.1.1.3.a. Rising rotating stem, globe valve.

Although the closure element in these valves moves in a linear fashion, the actuator output is rotary. The rotary motion is converted to a linear motion by the stem nut.

For valves with no linear to rotary conversion mechanism, a linear output actuator would be needed to provide the required motion. Many valves have smooth stems with no threads, hence there are many types of linear actuators. The most common are pneumatic linear actuators used for the operation of process control valves such as the one shown on the following page.

Figure 2.1.1.3.b.
Linear output, spring diaphragm pneumatic actuator on a process control valve. *Image courtesy of Richards Industries*.

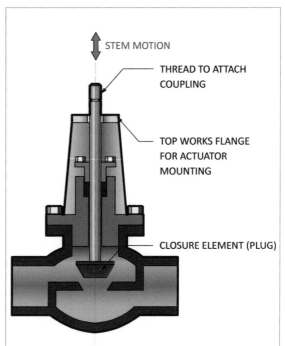

STEM MOTION

THREAD TO ATTACH COUPLING

TOP WORKS FLANGE FOR ACTUATOR MOUNTING

CLOSURE ELEMENT (PLUG)

Figure 2.1.1.3.c.
Sliding stem valve.

2.1.2   Valves with part-turn closure element motion

Part-turn valves are not a new invention although changes in design and materials have increased their usage in most industries.

The majority of part-turn valves require a quarter-turn of the valve stem to move from closed to open and visa-versa. This type of valve, having a standard movement of 90 degrees, allows an actuator with a known travel to be utilized. Unlike the multi-turn actuator, there is generally a fixed amount of travel on the quarter-turn device.

The closure element on the quarter-turn valve could be a disc (butterfly valve), a sphere (ball valve), the frustum of a cone (tapered plug), or a variation or combination of these types.

Figure 2.1.2.a. Metal-seated butterfly valve.
Image courtesy of Emerson Virgo.

Figure 2.1.2.b. Resilient seated butterfly valve. *Image courtesy of CIRCOR Energy.*

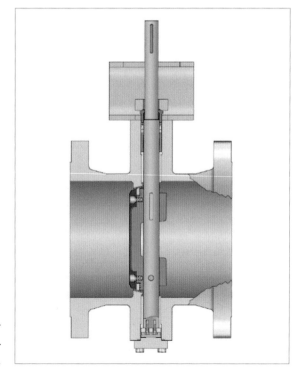

Figure 2.1.2.c. Butterfly valve, in section. *Image courtesy of Velan.*

Figure 2.1.2.d.
Ball valve.
*Image courtesy of CIRCOR
Energy.*

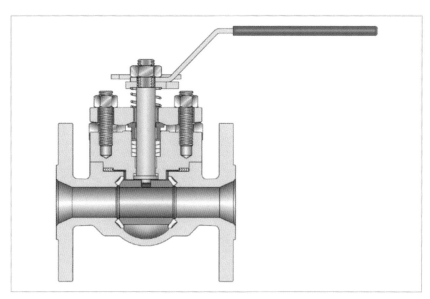

Figure 2.1.2.e. Ball valve, in section.
Image courtesy of Velan.

2.2   Force

The force required to move the valve through its range of motion is critical to the selection of actuator. In the case of sliding stem valves, this force would be linear and would be considered a pushing or a pulling force. For multi-turn or part-turn valves a rotary force or torque is required.

A linear force in the SI system is expressed in Newtons or in the imperial system in pounds force. Rotary force or torque is measured in the SI system in Newton meters or in the Imperial system in foot pounds or for smaller valves inch pounds.

Torque is defined as force times the perpendicular distance to the axis of rotation of the force. So an applied force of 10 pounds on a hand lever of 2 foot length would produce torque of 20 foot pounds.

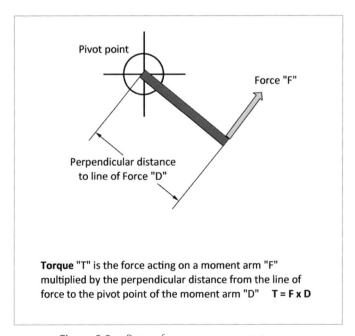

Figure 2.2.a. Rotary force measurement, torque.

The torque required to operate a valve usually varies with the size of the valve, the valve design and the differential pressure across the valve. The differential pressure is the difference between the upstream pressure on the valve and the residual downstream pressure. Usually this pressure difference is greatest when the valve is closed. Typically, it is this differential pressure that makes the force required to unseat and open the valve the greatest force requirement of the valve.

The friction between the valve closure element and the valve seats, with the applied differential pressure, causes the maximum force requirement of the valve. The larger the valve closure element, the larger the force on the seats. Similarly, the larger the differential pressure across the closure element, the larger the force on the seats.

For some valves, such as gate and globe valves, this force requirement can be calculated and a reasonable estimate of the valve force demand can be made for sizing the actuator. Although other considerations need to be included such as stem packing friction, the media temperature and the mechanical characteristics of the valve and valve stem. *See appendix 14.1 for details.*

For quarter-turn valves, however, it is not possible to independently calculate the torque demand of the valve with accuracy. This has to be done by physical measurement of each valve size under the various differential pressure conditions. The valve maker does this and provides the information for sizing actuators. *See appendix 14.1 for details.*

Valve power demand

Linear (class 300 gate)

		Size min Inches mm	Size max Inches mm	Nominal force Size min Pounds, N	Nominal force Size max Pounds, N	Speed normal Seconds	Speed fast Seconds	Nominal power HP , W Speed normal Size min	Speed normal Size max	speed fast Size min	speed fast Size max
Large	Imperial Units	8	24	11009	95455	60	30	0.219	5.694	0.438	11.388
	SI Units	200	600	48970	424605	60	30	163	4246	326	8492
Intermediate	Imperial Units	4	8	3091	11009	60	10	0.031	0.219	0.184	1.313
	SI Units	100	200	13749	48970	60	10	23	163	137	979
Small	Imperial Units	1	4	618	3091	30	6	0.003	0.061	0.015	0.307
	SI Units	25	100	2749	13749	30	6	2	46	11	229

Quarter-turn (class 300 ball)

		Size min Inches mm	Size max Inches mm	Nominal force Size min Inch Pds, Nm	Nominal force Size max Inch Pds, Nm	Speed normal Seconds	Speed fast Seconds	Nominal power HP , W Speed normal Size min	Speed normal Size max	speed fast Size min	speed fast Size max
Large	Imperial Units	8	24	11442	89722	120	30	0.023	0.178	0.091	0.712
	SI Units	200	600	1293	10139	120	30	17	133	68	531
Intermediate	Imperial Units	4	8	2918	11442	60	10	0.012	0.045	0.069	0.272
	SI Units	100	200	330	1293	60	10	9	34	52	203
Small	Imperial Units	1	4	1000	2918	30	6	0.008	0.023	0.040	0.116
	SI Units	25	100	113	329.734	30	6	6	17	30	86

Figure 2.2.b. Some typical valve power demands.

This table is a guideline for the range of power levels needed for a typical gate valve and a typical ball valve for both normal and fast operation. It gives a perspective on the power required at the output of an actuator. This relates to large, intermediate and small types of actuator described in section 6.

Valve sizes can range from micro-valves with orifices down to fractions of a square inch or square mm, to massive pipeline valves with orifices of several square feet or square meters. Similarly, the differential pressures could vary from very little to several thousand times atmospheric pressure.

The range of output force or torque needed to cover this spectrum of valves is spread over several types and designs of actuator. The greater the valve torque demand, the larger the actuator required and the more costly it is likely to be.

2.3   Speed

When sizing an actuator, the first consideration is the maximum force required by the valve. This determines the force output of the actuator. The speed of operation, when considered with the force required, defines the power needed. Power is defined as work done in a specific time. The work done by an actuator is the force demanded by the valve multiplied by the distance over which that force has to be applied, the valve travel. If this work has to be done over a period of time of, for example, one minute, then the power needed would be twice as much as that needed to operate the same valve over a period of two minutes.

For electric actuators this is a critical parameter as it dictates the motor power required. For fluid power actuators it impacts the size of the supply and exhaust lines as well as the sizing for the direction control valve. In both cases there is a significant impact on the cost, as well as physical size, of the actuator.

Another important consideration is inertia when high speed valve operation is required.

The high inertia generated, either by the motor on an electric actuator or the piston on a fluid power actuator, potentially could damage the automated valve assembly. The problem is resolved in fluid power and electric actuators by slowing the speed of the actuator as it approaches the valve seat. This can be done by the use of a snubber in a fluid power circuit for a fluid-powered actuator, and by motor speed control in electric actuators.

For valves and pipelines, a further consideration is the hydrodynamic shock or water hammer in the pipeline that may be generated by rapid closure of a valve. *For a fuller explanation, see appendix 14.5.*

2.4   Frequency of movement

The preceding valve demand criteria dictated the needed output motion, force output and power of the actuator. The required frequency of operation has a direct impact on the durability of the mechanical drive and robustness of the controller.

Isolating duty valves and actuators usually need only operate infrequently, perhaps only once or twice a day. With a properly sized actuator, there is little wear on mechanical components and controls.

Modulating valves, however, can operate constantly as they control a process. This requires a degree of resistance to wear on the valve and actuator assembly so it's important that modulating requirements are factored into the actuator selection. Not only is there greater mechanical wear on the drive train, but the controls need to be capable of constant change without overheating or failing.

There are many subjective opinions on the definition of modulating and regulating duty. However, the industry has adopted milestones of 60 and 1200 starts per hour for actuator capability. Figure 2.4.a., the frequency of operation chart attaches some general nomenclature to those start frequencies.

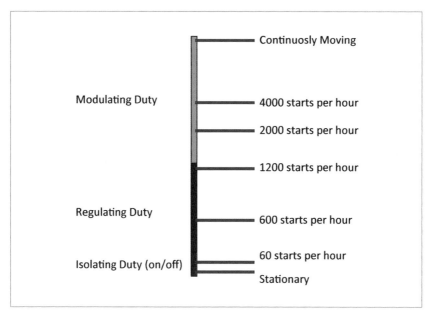

Figure 2.4.a. Frequency of operation chart.

Chapter 3

Power source

Every actuator needs some form of external power source to operate. Even manual gearboxes need human power, so the choice of actuator is strongly influenced by the power sources available.

Ever since the early days of valve actuator use, there has been a choice between powering the actuator with electricity or some type of pressurized fluid. Sometimes a user industry has a traditional preference for actuator power and sometimes it's dictated by the circumstances of the application.

Deciding the best power medium for an industrial application depends on many factors. Often it is simply a matter of choosing what has traditionally been used for the application. But new technology in actuators and power sources allow the traditional solutions to be more closely examined.

For plants that have shop or instrument air systems, the choice of actuator power has more flexibility. Because electric power is often used to power the air supply, either fluid or electric powered actuators could be used. But there are many installations where an electric power supply is not available, these are usually confined to remote sites like wellheads, pipeline sectioning valves or irrigation systems. Also there are applications where electricity is not used due to safety concerns.

In places where both power modes are available, then the choice comes down to other criteria.

Some industries have their own history on the power supply used. In the earlier days of the power industry, many boilers were controlled by instrument air controllers such as the Bailey Controller. Boiler valves were controlled by pneumatic controllers and positioners using spring diaphragm and piston actuators, mainly on the smaller valves. Larger valves such as the main steam stop valves became electrically actuated. Modern power plants typically use more electric actuators, but pneumatic modulating valves are still used extensively, although they are controlled and positioned electronically rather than pneumatically.

Some early designs of filter plants for the potable water industry used pressurized water as a power medium for piston actuators. These have been superseded by electric actuators in the majority of new plants around the world, but there are some plants being built using pneumatic actuators for filter control.

Oil and gas production has a long history of using pneumatic actuators offshore and onshore, probably because of the mechanical simplicity of piston operators and their straightforward maintenance. Also, an instrument air supply poses no sparking hazard to the explosive environment on an oil rig, although any electric control used must be hazardous-area rated.

Some oil and gas companies are using electric actuators offshore for certain valves, particularly where there are weight or space restrictions.

The functionality of a fluid-powered actuator has an advantage over the electric actuator in its capacity to provide a simple mechanical fail to position function. The cylinder or diaphragm of a fluid power actuator is easily opposed by a spring that can move the valve to an open or closed position on loss of power source.

For process shutdown or safety shutdown, the fluid power actuator is the traditional choice, particularly when high levels of safety and integrity are required. There are some technologies that can provide integral backup power in the form of batteries or super-capacitors for electric actuators. These are becoming a viable alternative to the traditional fail-to-position function for process valves but they are limited to smaller valves at present.

Gas pipelines often use the pressurized pipeline gas to power shutoff valves for line break sectioning and general shutdown. These fluid-powered units often use oil as an interface between the pipeline gas and the actuator. Hence the term "gas over oil" is used to describe the actuator. However, many countries are restricting the use of gas-powered actuators due to the environmental impact of the exhausted gas. One replacement solution is the electro-hydraulic actuator, which can provide shut-down capability using a spring return hydraulic actuator. For remote locations, these units can be solar powered, using a DC motor powered hydraulic pump to drive the reset stroke after a shutdown.

For smaller process control valves, the traditional actuator is the pneumatic spring diaphragm unit. This is a simple but effective device with few moving parts. To make a complete automated valve, it is usually coupled to a positioner, though nowadays it is often a "smart" device. Most process control valve makers manufacture their own diaphragm actuators and package them with the valve and positioner as a complete unit.

The relative capital cost of electric actuators when compared to fluid power units of equivalent output is usually greater, although the electro-hydraulic designs are the exception as they combine components of electric and fluid power units.

The operating cost of actuators does not always figure prominently in evaluations, but for frequently modulating valves, this could be significant. Electric actuators look good in these scenarios as they only use significant amounts of energy when moving. Fluid power units using instrument air have a constant energy draw from positioner bleed and system leakage.

3.1   Electric

The most common primary power source for actuators is electricity, but not just for electric actuators. Electricity is often the indirect source of power for fluid power actuators. Instrument air and central hydraulic power units are often powered by an electrically driven compressor or pump. In those circumstances the choice of using fluid power actuators rather than direct electric actuators is predicated on other requirements and considerations.

If there is easy access to an electrical supply, then the valve actuator can be powered directly using an electric motor. Higher powered actuators typically use a 3-phase supply, smaller actuators can be powered using a 1-phase supply or even a DC supply. The power delivered by an electric supply is the product of voltage and current.

> For 3-phase supplies:
> Power (watts) = √3 x PF (Power factor) x volts x amps
> For 1-phase supplies:
> Power (watts) = PF x volts x amps
> For DC supplies:
> Power (watts) = volts x amps

For larger valves and ones with greater differential pressure, more power is required to automate them. That is why 3-phase power is so often used in these circumstances. This applies to both direct electrically automated valves as well as fluid-powered valves powered by a motor-driven compressor or pump. As a general rule, when valve sizes are larger than about 4" to 6" (100mm to 150mm) and depending on pressure class, then 3-phase is often needed.

For smaller valves, where the power demand is less, then 1-phase electricity is often used. However, 1-phase power has a proportionately lower voltage than a 3-phase supply. This means that the current increases.

For example if we had a 6" valve that needed 50 watts to operate in 30 seconds.

The current in amps needed from a 3-phase 480 volts supply would be derived

> Power (50) = √3 x PF (say 0.8) x 480 volts x amps
> So the amps needed would be 50/(1.732 x 0.8 x 480)
> = 0.075 amps

For a 1-phase supply of 110 volts

> Power (50) = PF (0.8) x 110 volts x amps
> Amps needed would be 50/(0.8 x 110) = 0.568 amps

This means a 1-phase supply for a given valve application requires nearly eight times the current.

The offset for this greater current requirement is the cost of the cable and insulation required for the higher voltage 3-phase supply.

If we were to look at using a 24 volts DC supply for the same application, the amps needed would be:

$$\text{Amps} = 50 \text{ watts} / 24 \text{ volts} = \underline{2.08} \text{ amps}$$

This is why low-voltage actuators are best confined to smaller valves.

Where there is no readily available power supply, such as remote drill sites or pipelines, then power has to be taken either from the pipeline media, a rechargeable gas cylinder or from the sun.

The most common energy source for actuators on gas pipelines is the gas pressure inside the pipeline. However, this has its problems. To use the energy in the gas pressure, it is expanded and then vented to the atmosphere. This means that not only is the valuable gas lost, but the environment is polluted by the gas.

For these reasons, in environmentally sensitive countries, the industry is looking more to solar power for remote locations. But as we have shown from the above examples, this is only feasible for lower-powered actuators.

The exceptions are large shut-down actuators that can be powered by a low-voltage DC solar supply. This is achieved by opening the valve slowly and at the same time compressing a spring. This uses less power because the work is done over an extended period of time. Under emergency shut-down conditions, the energy stored in the spring is then used to rapidly shut the valve. So this type of actuator can be considered low power for opening, but delivers a high-powered spring closing stroke.

3.2   Fluid power

Pneumatic- and hydraulic-powered actuators require an infrastructure to provide their power. For pneumatic actuators, this would take the form of an instrument air supply.

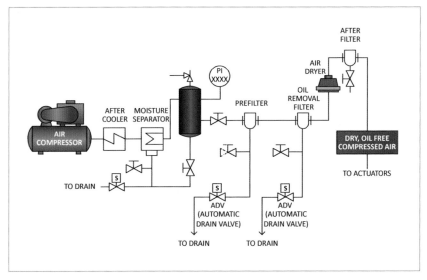

Figure 3.2.a. Instrument air supply.

The air compressor is usually driven by an electric motor, but a combustion engine could be used; or, in a steam cycle power plant for example, it could be steam turbine driven. The pneumatic actuators are connected to the compressor installation by supply lines and the expended air is vented to the atmosphere. An instrument air supply for an actuator could be derived from a "shop" air supply. Shop air is often produced to a lower standard of moisture and particulate content. To upgrade shop air to instrument quality would require additional filtration and drying.

General specification for instrument air		
Particle size	Dewpoint	Lubricant content
< 3 to 4 microns	10C below lowest expected ambient	<1 ppm

Hydraulic actuators can draw hydraulic power from a central hydraulic power unit (CHPU). This is usually an electric-driven hydraulic pump with associated equipment such as hydraulic accumulators for fluid power energy storage as well as filtration and other peripherals. The hydraulic actuators are connected to the CHPU by high-pressure supply and low-pressure return lines.

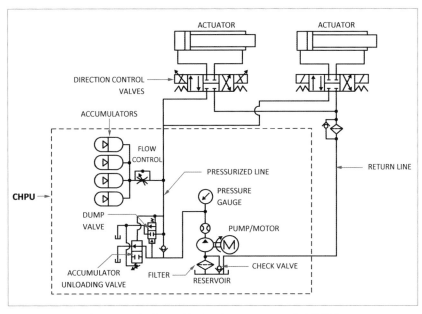

Figure 3.2.b Central hydraulic power supply (CHPU).

The selection of power supply may depend on the availability of alternatives. Selection is also a function of the size of the valves to be automated; larger valves need more power.

The other consideration is the function of the valves. If the valves are needed to provide a shut-down function for safety or process requirements, then fluid-powered actuators are often chosen. This is because energy can easily be stored in the actuator by the use of a compressed spring. (This is also possible in electric actuators, but the spring drive mechanisms are more complex). In addition, fluid power can be stored in hydraulic or pneumatic accumulators to provide one or more strokes of valve operation should the main power source fail.

An alternative is to utilize a self-contained electro-hydraulic actuator configuration. These are hydraulic actuators that have their own integrated hydraulic power supply. The hydraulic pump is driven by an electric motor on the actuator. The motor can be powered using an AC or a DC electric power source, so solar power is a possibility. The solar power pack then allows the actuator to be used on shut-down service for valves in remote locations.

Chapter 4

Working environment

Valves can be found in every type of environment on the planet. A valve is designed not only for exposure internally to the pipeline media, but also to handle the surrounding environment.

Valves are robust devices with relatively few vulnerable parts. Actuators, however, contain controls and instrumentation that could be adversely affected by extremes of humidity, temperature and environmental ingress. For that reason there is a range of alternative types of enclosures that can be incorporated in the actuator design to protect these vulnerable parts.

The selection of the appropriate enclosure design is dependent on the expected environment. The user or specifier of the automated valve must determine what environment is expected for the assembly so the proper enclosure is supplied. In some applications, particularly in the oil and gas industries, the atmosphere may be potentially explosive. The appropriate hazardous area certified enclosure must be used for the actuator and controls.

Key considerations are:

1) Temperature range
2) Enclosure environmental protection rating
3) Hazardous area rating (explosion proofing)

4.1 Temperature

If we assume a new outside installation is to be designed and built using automated valves, then for that specific area, the historic meteorological data will show the temperature range as well as the precipitation. The hazardous area rating will be determined by the designing engineers after considering the process details.

Valve actuators are often the most vulnerable part of an automated valve assembly. Electric actuators contain a mixture of electric, electronic, elastomeric and metallic components. Fluid-powered actuators, though not directly electrically powered, will contain similar materials for control and feedback components.

For most installations in industrialized countries, the standard valve actuator is designed to accommodate a common temperature range of about -22F to 158F (approximately -30C to 70C).

For installations where there are extremes of temperature, enhanced temperature capability will be required for the actuator.

In the artic regions of Canada, Alaska or Russia, some engineering specifications require operation at temperatures as low as -76F (-60C). Conversely, on the high temperature side, some Arabian Gulf countries require operation at temperatures as high as 176F (80C). Clearly these extremes are generally outside the typical standard range of even heavy-duty actuators. Special modifications, materials and even design changes are needed to accommodate these extremes. Many manufacturers can accommodate the challenges provided the requirement is clearly specified.

Elastomers for sealing fluid power cylinders and other enclosures must retain resilience to form an effective seal. At low temperatures some elastomers become too stiff, and leaks occur. At high temperatures the seal must neither melt nor harden by vulcanization. Special seal materials such as Viton are often chosen for extremes of temperature rather than the commonly used Nitrile rubber. Care must be taken when selecting lubricant used to ensure compatibility with the seal material to avoid swelling or other deterioration of the seal.

For cold temperatures there is often a misunderstanding regarding the "wind chill factor." Wind chill describes the rate of heat transfer, and although it's of significant impact to living creatures, it has little or no bearing on the temperature rating for valve actuators.

4.2   Enclosure protection rating

Precipitation, humidity and the resulting moisture is the enemy of all valve actuators. The electric, electronic and position-sensing components are vulnerable to moisture exposure and the subsequent corrosion. Some sites see extremes of precipitation, storms and flooding. Actuators may need to withstand torrential rain and humidity even to the point of temporary submersion.

The sealing capability of electrical enclosures can be defined using a rating convention such as ingress protection (IP) or National Electrical Manufacturers Association (NEMA). Actuators and their assembled control accessories need to be carefully specified to properly ensure their suitability for the intended working environment.

The differing environments encountered by automated valves can be designated using the IP/ NEMA rating table:

NEMA Ratings and IP Equivalency Chart				
NEMA Rating	IP Equivalent	NEMA Definition	IP Definition	
1	IP10	Enclosures for indoor use, some protection against incidental contact with the enclosed equipment and some protection against falling dirt	1 = Protected against solid foreign objects of 50mm in diameter and greater	0 = Not Protected
2	IP11	Enclosures for indoor use, some protection against incidental contact with the enclosed equipment and some protection against falling dirt and dripping and light splashing of liquids	1 = Protected against solid foreign objects of 50mm in diameter and greater	1 = Protected against vertically falling water drops
3	IP54	Enclosures for either indoor or outdoor used to provide some protection to personnel against incidental contact with the enclosed equipment, against falling dirt, rain, sleet, snow, windblown dust and ice formation.	5 = Protected against dust - Limited to ingress (no harmful deposit)	4 = Protected against water sprayed from all directions - Limited to ingress permitted.

NEMA Rating	IP Equiv-alent	NEMA Definition	IP Definition	
3R	IP14	Enclosures for either indoor or outdoor used to provide some protection to personnel against incidental contact with the enclosed equipment, against falling dirt, rain, sleet, snow, and ice formation.	1 = Protected against verti-cally falling water drops	4 = Protected against water sprayed from all di-rections - Limited to ingress permitted.
3S	IP54	Enclosures for either indoor or outdoor used to provide some protection to personnel against incidental contact with the enclosed equipment, against falling dirt, rain, sleet, snow, windblown dust and operable when ice laden.	5 = Protected against dust - Limited to ingress (no harmful deposit)	4 = Protected against water sprayed from all di-rections - Limited to ingress permitted.
4	IP66	Enclosures for either indoor or outdoor used to provide some protection to per-sonnel against incidental contact with the enclosed equipment, against falling dirt, rain, sleet, snow, windblown dust, splashing and hose directed water. Undamaged by ice formation.	6 = Totally pro-tected against dust	6 = Protected against strong jets of water from all di-rections - Limited to ingress permitted.
4X	IP66	Enclosures for either indoor or outdoor used to provide some protection to personnel against incidental contact with the enclosed equipment, against falling dirt, rain, sleet, snow, windblown dust, splashing and hose directed water and corrosion. Undamaged by ice formation.	6 = Totally pro-tected against dust	6 = Protected against strong jets of water from all di-rections - Limited to ingress permitted.
5	IP52	Enclosures for indoor use to provide some protection to personnel against incidental contact with the enclosed equipment; to provide some protection against falling dirt, airborne dust, lint, fibers, and a degree of protection against dripping and light splash-ing of liquids.	5 = Protected against dust - Limited to ingress (no harmful deposit)	2 = Protected against direct sprays of water up to 15° from the vertical.
6	IP67	Enclosures for either indoor or outdoor use to provide protection to personnel against incidental contact with the enclosed equip-ment; to provide some protection against falling dirt; against hose-directed water and the entry of water during occasional tempo-rary submersion at a limited depth; undam-aged by the external formation of ice on the enclosure.	6 = Totally pro-tected against dust	7 = Protected against the effects of temporary im-mersion between 15cm and 1m. Duration of test 30 minutes.
6P	IP67	As 6 , but prolonged submersion	6 = Totally pro-tected against dust	7 = Protected against the effects of temporary im-mersion between 15cm and 1m. Duration of test 30 minutes.

Figure 4.2.a. Table of IP and NEMA enclosure ratings.

4.3   Hazardous area rating

Many sites in the oil and gas industry have combustible gases in the atmosphere that could lead to fire or explosion if ignited. To ensure that the electrical equipment inside the valve actuator does not allow a spark to ignite this atmosphere, the electrical components are housed in an enclosure that is strong enough to contain an internal explosion. The assumption is that explosive gases may find their way into the electrical enclosure and when switch contacts make or break, the spark generated could ignite the gases in the actuator. The enclosure must not only contain the explosion but also quench any flame that may escape through the cover joints. To achieve this quenching, the enclosures are designed with a flame path of a required gap and length such that any escaping flame is rapidly cooled, and therefore quenched, as it passes along this path.

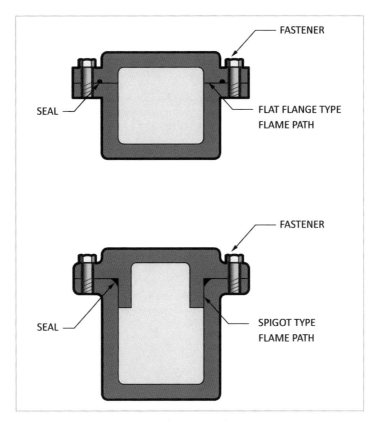

Figure 4.3.a. Types of enclosure for hazardous areas.

This method of hazardous area protection is one of containment.

An alternative method is to limit the power supplied to the components to a level low enough to preclude ignition (intrinsic safety). This means limiting voltage and power. This method of protection is predominantly applied to low powered instruments or actuators. Occasionally, larger powered actuators employ a hybrid system that uses containment on the motor and intrinsic safety for the controls.

To ensure safety with electrical equipment, there are independent certifying bodies that will examine and test the suitability of equipment for use in hazardous locations for a manufacturer. Unfortunately, there is no universally accepted global standard for this certification, so local compliance is required. For example, CSA certification is required in Canada, FM or UL is needed in the USA, IEC is required in Australia, GOST is used in Russia and other certifications are required in other countries. *A list of the applicable certifying bodies can be found in Appendix 14.3.*

Chapter 5

Actuator controls

5.1 Force/ torque sensing

Valve actuators are devices for delivering either linear or rotary force. It is important that the amount of force delivered to a valve is sufficient to operate it effectively, but not so great it will damage the valve. Valves are mechanical devices with seats and stems that can be damaged by excessive force. *See appendix 14.2 valve stem buckling load.*

For that reason, almost all electric actuators have some form of force-limiting mechanism. The exceptions are some of the smaller electric or fluid power actuators that produce relatively low maximum forces.

A large portion of fluid-powered automated valves, especially pneumatic, are on part-turn valves such as ball and butterfly valves where the sealing of the valve is effected by virtue of the position of the ball or disc. The force of the actuator is contained in the end stops of the actuator so it need not be transmitted to the valve to achieve sealing.

On linear applications the force output of the fluid power actuator is limited by the power supply system pressure. In almost all cases this system pressure is controlled by a pressure regulator on or adjacent to the actuator. The maximum allowable force on the valve is often several times the force required to

Figure 5.1.a. A sluice gate with a long stem, vulnerable to buckling if overloaded.
Image by permission of Rotork®.

operate the valve, so lack of precise force control on fluid-powered actuators is often not a problem.

This is not the case with electric actuators. Early applications for electric actuators were often on wedge gate valves where the seating force had to be carefully controlled. From its earliest designs the electric actuator has had an internal torque limiting mechanism. This is to ensure sufficient force for seating and sealing the valve, but prevent excessive force that could gall the seats or cause the valve to jam.

There are many ingenious methods of measuring the force output of an actuator. Most of these take advantage of a measurement of the reaction force exerted at some convenient point in the motorized gear train.

For worm and wheel gears, the axial force reaction of the worm shaft against the worm wheel is in direct proportion to the output torque of the actuator. This is measured by monitoring the movement of the worm shaft against a spring. Alternatively, some devices use an electronic pressure sensor to measure the force reaction on the worm shaft.

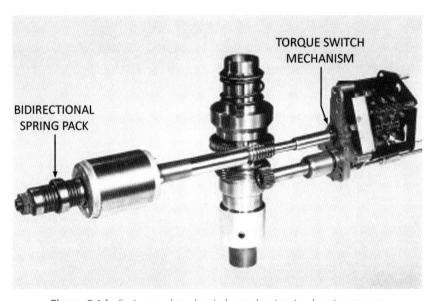

Figure 5.1.b. Spring pack and switch mechanism in electric actuator.
Image by permission of Rotork®.

THRUST SENSOR

Figure 5.1.c. Piezoelectric thrust sensing assembly.
Image by permission of Rotork®.

On epicyclic gears the torque reaction on the outer gear ring can be measured to gauge the output of the actuator. Where the outer ring gear also acts as the worm wheel for the manual override function, the measurement of the axial force on the manual override worm shaft gives the torque measurement.

Electric motor current measurement also can indicate the torque that is being generated. However, other parameters such as temperature need to be measured in order to calculate the torque generated using a calibration of torque/current for the given motor. Motor speed can give an indication of torque by interpolation of a known torque/speed curve for the motor.

These and other methods have been incorporated into the design of the actuator to provide a mechanical or electronic output, which in turn controls the force output of the actuator. Once the predetermined force or torque limit is reached, a torque sensor is tripped and power is shut off to the motor.

The predetermined force can be adjusted so that one size of actuator can be used on different valves or applications. The sizing process for the valve will give the configuring technician the data to set up the actuator to suit the valve.

5.2 Position-sensing

Position-sensing is important for two reasons:

1) To give feedback on the position of the valve. This could be transmitted to the remote control room or displayed locally on the actuator.
2) To allow the valve to be seated correctly where applicable.

Some valve designs require a specific amount of force to seat the closure element sufficiently so that the pipeline medium cannot pass. These valves are typically wedge gate, globe or triple offset butterfly valves.

Other valves, including most quarter-turn valves such as ball, plug and resilient seated butterfly valves, plus some slab or knife gate valves, are designed to seal at a certain position.

To ensure proper seating on these "position-seating" valves, the actuator must move to the correct position and stop. One way to achieve this is to have a mechanical stop in the actuator. This is often provided on electric or fluid power actuators for quarter-turn valves, but position-sensing is still needed for indication. It also often is achieved by a direct drive from the part-turn actuator output shaft to a switch trip mechanism such as cams or levers. In addition, a potentiometer or a 4-20 milliamp transmitter can be driven from the same shaft to give continuous remote position indication.

For operational purposes the actuator must always know where it is, relative to the valve position. For that reason a position measuring system is mandatory in multi-turn actuators for "position-seating" valves. A mechanism driven from the output of the actuator achieves this.. This mechanism can be used to stop the actuator at each end of travel, either directly by a switch trip mechanism or indirectly by a processed electronic measurement to control the motor. The multi-turn actuator needs a versatile counting mechanism to accommodate the variety of output turns required by different valves. There is usually a shaft driven from the actuator output that connects to a mechanism for counting turns.

Some mechanisms count the output turns of the actuator using a geared rotating counter-type mechanism, similar to an odometer mechanism but bi-directional. Another method uses a rotating threaded shaft on which a nut travels in proportion to the actuator output turns. Both types of mechanical drive mechanisms have a finite number of output turns that they can accommodate.

In addition, these mechanisms have a means of driving a continuous position measuring device. In past years this device was a part-turn or multi-turn potentiometer. For valves requiring many actuator output turns, there had to be additional reduction gearing on the potentiometer drives to accommodate the extended travel. This made the device more complex, harder to set up and harder to specify at time of order.

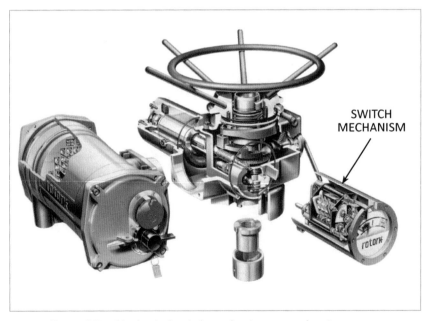

SWITCH
MECHANISM

Figure 5.2.a. Mechanical switch mechanism on an electric actuator.
Image by permission of Rotork®.

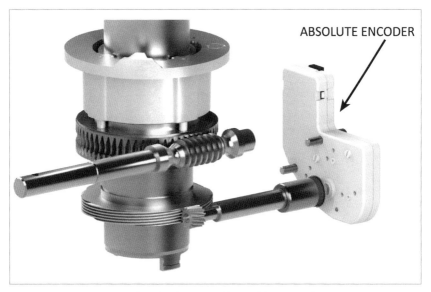

Figure 5.2.b. Absolute encoder and drive mechanism.
Image by permission of Rotork®.

The more common method now is to use an electronic encoder, which can measure a much greater span of output turns. This has become the common method of position-sensing in smart actuators.

There are two types of encoders in general use for valve actuators: the absolute and the incremental. The incremental encoder will count the number of turns from a set position established in its memory during the set-up procedure. It can only count when powered. If the incremental encoder is powered down and the actuator moves, then it loses its position reference. Because of this, a backup battery is usually incorporated in the design to power the encoder and its processor, should the actuator lose power.

The absolute encoder works differently. It does not need to be powered when moved as each position has a unique signature that can be read at any time. It does not rely on an incremental count to maintain its position reference.

However, it is beneficial for actuators with either type of encoder to have a battery backup supply to ensure position indication as well as to operate position indicating switches and relays should the actuator lose power.

The position-sensing mechanism is used to trip the motor power at the ends of valve travel on position-seating valves. It is used on all valves to give position information to local operators and the control room. Where sequences are part of a process, the switches are frequently used to give confirmation that one valve is in the correct position before another is moved.

The continuous position-sensing function is essential in process control valve assemblies. The valve positioner needs to know exactly where the valve is so it can determine its required direction of movement relative to the process variable (flow, pressure, temperature or other parameter). It can then move the valve as required to maintain the desired process conditions.

5.3   Direction control

All types of actuators need some kind of direction control to move the valve in the opening or closing direction.

For electric actuators this usually takes the form of a motor starter. For fluid-powered actuators a direction control valve is used.

Figure 5.3.a. Fluid power direction control valve with solenoid.
Image by permission of Rotork®.

Figure 5.3.b. Electric motor reversing starters for a 3-phase motor.
Image by permission of Rotork®.

In both cases an external control signal is used to energize the coils of the motor starter or the solenoid coils of the direction control valve.

Control wires connect the control room to these directional controls and are often 110 volts AC or 24 volts DC.

For fluid-powered actuators the power supply is clearly a separate medium from the control circuitry.

High-powered electric actuators usually use a 3-phase supply.

For electric actuators an important decision needs to be made on the control of the motor. There are two main types of motor control layout:

1) A separate motor control center (MCC) that contains the motor starters for the actuators in one central location separate from the valves. This configuration is often used when actuators are located in hostile environments.
2) Motor starters integral to the actuator located at the valve. This configuration provides a simpler and often less expensive installation.

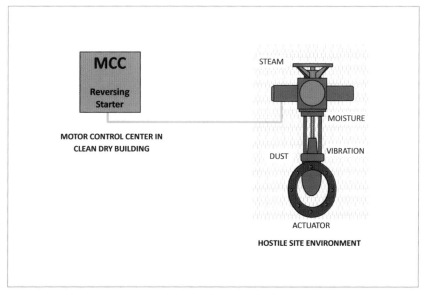

Figure 5.3.c. Actuator located separately from motor starter.

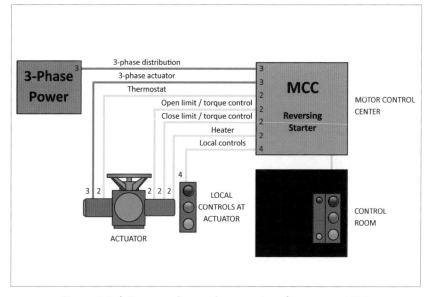

Figure 5.3.d. Power and control connections for separate MCC.

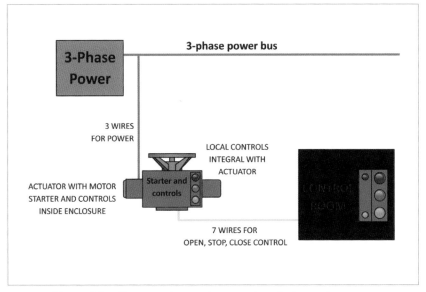

Figure 5.3.e. Power and control connections for integral starters .

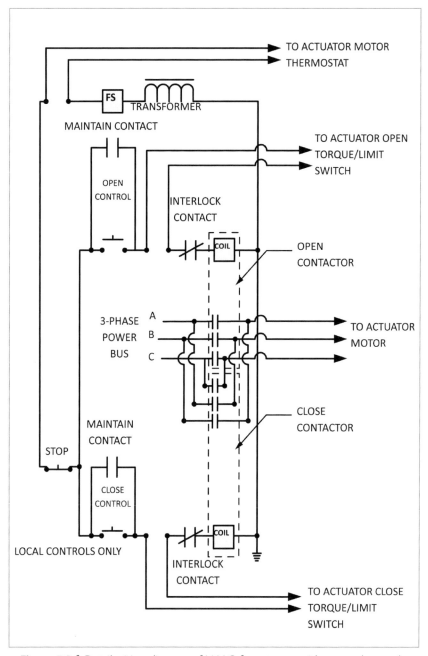

TO ACTUATOR MOTOR

THERMOSTAT

FS

TRANSFORMER

MAINTAIN CONTACT

OPEN
CONTROL

INTERLOCK
CONTACT

TO ACTUATOR OPEN
TORQUE/LIMIT
SWITCH

COIL

OPEN
CONTACTOR

3-PHASE
POWER
BUS

A

B

C

TO ACTUATOR
MOTOR

CLOSE
CONTACTOR

MAINTAIN
CONTACT

STOP

CLOSE
CONTROL

LOCAL CONTROLS ONLY

COIL

INTERLOCK
CONTACT

TO ACTUATOR CLOSE
TORQUE/LIMIT
SWITCH

Figure 5.3.f. Detail wiring diagram of M.M.C. for actuator with external controls.

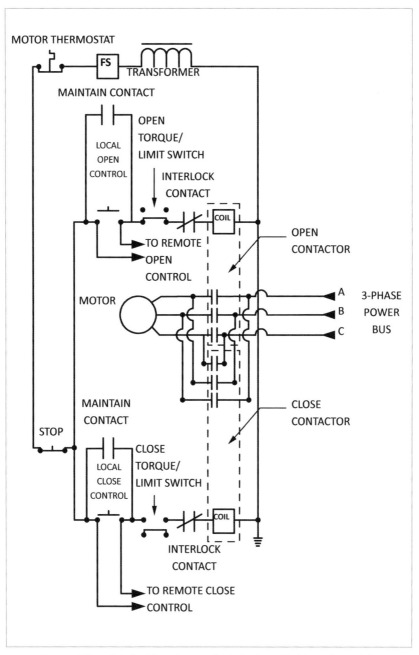

Figure 5.3.g. Detail wiring diagram for actuator with integral controls.

The separate MCC probably originated in the power industry where motor starters are typically grouped together. These include the starters for the boiler feed pump, the ID and FD fans as well as smaller equipment such as valve actuators. The control wiring runs from the control room to the MCC with feedback wires from the valve actuator position indicating and torque-sensing switches. The power cables run from the MCC to the individual actuators.

Originally the reason for the MCC layout was to keep the electrical control equipment in a clean, dry environment with a reasonable ambient temperature. With the advent of better environmentally sealed enclosures, the motor starters could be mounted within the actuator enclosure.

There are several advantages to the integral control method:

1) Torque and limit switches can be wired in the actuator, allowing the unit to be self-contained and factory tested.
2) There is less site wiring required.
3) Automatic phase correction can be incorporated in the controls. *(see appendix 14.4).*
4) A digital field bus link can be employed to reduce the multiple control and indication wires and also bring back diagnostic information.

The advantages of the MCC layout for actuator control are:

1) Motor starters are all located in a single area for easy maintenance.
2) The motor controls are removed from vibration, steam, dirt, water and other contaminants.
3) The actuator requires a smaller space envelope and has less weight.

Both methods of control are commonly used. The MCC-type layout was the global standard in 20th century power plant design and is still seen in water treatment facilities and power plants of German design. In contrast, integral controls and starters are preferred in the oil, gas, petrochemical and most other industries.

5.4   External and remote control requirements

An important function of an automated valve is its ability to be controlled from a distance. This allows a plant or process to be controlled from a central location. The location could be a control room somewhere in the plant or a control center some distance from the plant or even in another country.

Many automated valves are designed to be in either the fully open or the fully closed position. These require relatively simple controls to move the valve to one or the other of these two positions. The signal from the control room can be transmitted to the actuator direction control (motor starter or direction control valve [DCV]) over control wires to order the actuator to move to the fully open or fully closed position. This is sometimes referred to as "on/off" or "isolating" service.

Once the automated valve has moved to the new positon, advising the control room that the move has been successfully completed may be required. To provide that feedback, the actuator will have position-indicating switches that will activate at each end of valve travel. A signal is then transmitted back to the control room via dedicated wires.

Figure 5.4.a. Position-indicating switch box for fluid-powered actuators.
Image by permission of Rotork®.

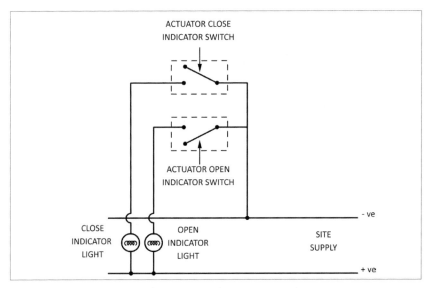

Figure 5.4.b. Position indicating wiring diagram.

With the advent of smart controls, much more information can be gathered from the automated valve, whether it's driven by a pneumatic-, electric- or hydraulically-powered actuator. For example, the valve force demand can be monitored relative to the valve position. This information can be used for predictive maintenance needs. The amount of data to be transmitted can be very large, so predictive maintenance data collection at the control room only became practical with the advent of digital field communications, which can carry high volumes of data over just a pair of wires.

5.4.1   Local controls

In many instances an application may require a method of locally controlling the automated valve. A local push-button station with a selector for "Local" or "Remote" control is provided adjacent to the actuator. For electric actuators that have integral controls, these functions are often incorporated into the body of the unit. The controls are packaged in the protected enclosure and are prewired by the manufacturer. This eliminates the additional cost of a separate control station, which may also have to be certified for the environmental conditions up to and including the hazardous location certification.

Figure 5.4.1.a. Actuator with integral motor starters and local push button controls. *Image courtesy of Auma.*

Figure 5.4.1.b. Motor starters and local push-button controls wall mounted for convenience and separate from the electric actuator. This configuration is also a method of isolating the actuator controls from vibration, temperature or other hostile elements at the valve. *Image courtesy of Auma.*

5.4.2 Positioning control

Some valves have to move more frequently than just the occasional open-to-close cycle that isolating valves perform. The valve may need to be moved to an intermediate position to regulate the flow of media in the pipeline. This requires a more sophisticated positioning capability. The control system has to send a signal to the actuator to tell it to go to the desired position, which is usually expressed as a percentage of the fully open position. The actuator may also need to confirm this positon back to the control room. This means the actuator has to have some means of measuring its intermediate position. This measuring device could be in the form of a potentiometer or a digital encoder mechanically coupled the output of the actuator.

For applications where the intermediate position is not frequently changed, the design of actuator could be similar to the isolating-type actuator. For example, the level control of a small lake could be controlled by movements of a valve every few hours. Conversely, if the valve is controlling the level of a small tank of liquid, then adjustments might be needed every few seconds.

There is no universal definition for the "frequency of operation" terms for electric actuators. The general rule is that if the actuator is capable of sustaining movement at a rate of more than 60 times in one hour, then it is considered a regulating device. At a rate of more than 1,200 starts in one hour, it's considered a modulating device and suitable for process control.

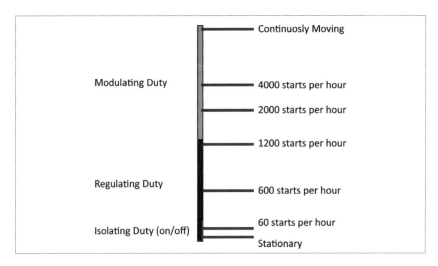

5.4.3   Modulating control

In the early 20th century, the primary method of controlling process control valves was by varying the pressure of a control air signal. Typically, this air pressure varied between 3 and 15 psi. A closed valve position would relate to 3 psi and the open valve position would relate to 15 psi.

Control valves were positioned by balancing this air pressure against an opposing spring. The higher the pressure, the more compression was exerted on the spring, and the greater the movement of the control valve. As the pressure backed off down to 3 psi, the spring force pushed the valve stem back to the original position.

This simple means of position control was used in a wide variety of process control applications and industries. It was the standard solution adopted by control valve actuator manufacturers as well as control system suppliers.

In its simplest form, compressed air was both the power medium and the control medium. Desired positions were achieved by varying this applied pressure and entire plants were controlled by compressed air channeled through small-bore copper tubing. The backs of control panels were a mass of tubes arranged by control systems craftsmen into symmetrical layers of carefully laid instrument pipe.

However, with the advent of computers and programmable logic controllers (PLCs) the days of the 3-15 psi control signal were numbered. Soon, they were replaced by electrical milliamp current signals carried on much lighter duty copper wire.

For fluid-powered actuators, the device used to translate or relay the applied control signal of 4-20 milliamps to air pressure acting on the diaphragm or piston of an actuator, is the electro-pneumatic valve positioner.

The simple pneumatic positioner has evolved, from the basic functionality of controlling applied air pressure using a low-pressure signal, into the smart positioner of today. It not only directs high-pressure air to the valve actuator, but also gathers information on pressure at the various positions within the actuator assembly to provide diagnostic information. This diagnostic information can be transmitted back over the 4-20 milliamp connection using a communications protocol such as HART.

5.5   Manual control or override

In many circumstances there may be a need to operate a valve manually. This could occur if the power were lost to the actuator, in an emergency, during maintenance or for other reasons.

Although many actuators have a manual override as a standard feature, for others it's an option and for still others it's just not available.

Large electric actuators invariably have a built-in manual override mechanism. This is usually facilitated by a clutch mechanism that will disengage the motor drive and then engage the hand drive. This design feature is included for safety reasons. If it was not there and the motor was inadvertently energized during hand operation, the handwheel could turn rapidly, causing a serious hazard to the operator. There are many design variations of the clutch mechanism. Almost all manufacturers employ a power preference whereby should the actuator be left with the clutch in the manual position, starting the motor would automatically disengage hand drive and engage the motor drive. For this reason lock out and tag procedures on site require electrical isolation on actuators during maintenance. Large actuator designs also have the facility to mechanically lock the clutch mechanism in manual override mode as required.

Figure 5.5.a. Clutch mechanism on output shaft of electric actuator.
Image by permission of Rotork®

The handwheel drive may be a direct drive to the output of the actuator or it could incorporate intermediate gearing. This depends on the force required to turn the output stem nut or drive bush. There are many guidelines and standards that call for a maximum force requirement to operate the handwheel. For example, a maximum rim pull of 80 lbs. or 36 kg is often stated in specifications.

To achieve this on larger actuators, the hand drive either utilizes the output worm gear set of the motor drive or has its own independent gear set such as a worm and wheel or bevel gear.

Some large or high-pressure valves require a great number of handwheel turns to move the valve from open to close. The mechanical advantage (force multiplier) of the handwheel may be very high, and in some cases the user may not be able to feel when the valve has reached its end of travel. This poses a real danger of over-stressing the valve and causing galling on the valve seats. On sluice gates the stems are very long; some applications are below grade level and have extended stems. Should excessive force be applied, a buckled stem could result. For that reason the valve position should be carefully monitored during manual operation. *See appendix 14.2 Valve stem buckling load.*

Intermediate- and small-size electric actuators sometimes use a smooth rimmed handwheel directly coupled to the motor. This simple manual option has the handwheel rotating when the motor turns, but because torques are low and the wheel is smooth, there is perceived to be little hazard.

Epicyclic gear drives often use the outer ring gear for manual operation. The outside is machined to accommodate a worm gear that acts as an anchor during motor drive, and when rotated, acts as the manual operator. This method requires no additional clutch mechanism so it has an elegant simplicity.

For some intermediate and small quarter-turn actuators, a worm and quadrant is used as the final drive. A clutch mechanism at this stage of the gearing allows a handwheel to drive the final worm, providing ease of use via the mechanical advantage of the worm set.

For most fluid power actuators, manual operation is an optional feature. The need for a manual operator is reduced when the actuator has a positive fail-to-position mode on loss of power. Often there is a local manual control on the DCV, which allows a local maintenance operative to move the valve to a different position.

Figure 5.5.b. Spring return pneumatic actuator with
hydraulic manual override (red handle).
Image by permission of Rotork®.

However, a manual override may be required sometimes for either double-acting or spring return actuators. On large fluid power actuators (pneumatic or hydraulic), these are almost always self-contained hydraulic devices. A hydraulic cylinder is connected to the actuator piston rod and is pressurized in the desired direction by either a reciprocating pump or rotary gear pump. The direction of movement of the actuator and valve is selected by the pumps own dedicated DCV or by the direction of rotation on the rotary pump handle.

On intermediate and small scotch yoke pneumatic actuators, a jackscrew is sometimes an option. This usually comprises a stainless steel acme threaded shaft that penetrates the outboard side of the pneumatic cylinder to push on the piston rod inside. For double-acting actuators an opposing jackscrew is mounted on the other side of the body. The penetration into the cylinder is sealed with a thread seal and a locking nut. Due to the limited efficiency of the acme thread, there is a size limit to this override method.

Jackscrews that use standard thread forms or materials other than stainless steel should be avoided.

Small pneumatic actuators do not often require manual operators. They are often wrench operated by simply applying a torque to the top or the bottom of the output shaft. This usually has opposing flats to accommodate a wrench.

Alternatively, a declutching worm gear operator can be mounted between the actuator and the valve adaption. This is a special gear set that allows the worm shaft to disengage the worm quadrant by means of a scroll mechanism.

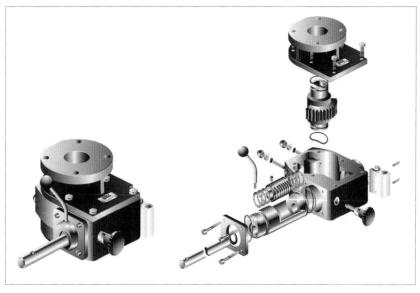

Figure 5.5.c. Manual override gear operator, quarter-turn with declutching worm shaft. *Image by permission of Rotork®.*

Chapter 6

Actuator types

— ◆ —

n chapter 2 we looked at actuators in terms of what characteristics affect the valve, that is force, motion, speed and frequency of operation. For a given valve, the requirements these characteristics create could be satisfied by an actuator powered by an electric, hydraulic or pneumatic power source.

As a way to catalog the types of valve actuators available, let's look at the logical selection process an engineer might employ to choose the right actuator mechanism for a valve application.

Once the valve characteristics are known, then the first defining characteristic of an actuator would be its power source.

The second defining factor could be the size needed to match the valve force demand, as this has the most impact on the overall power and size of the actuator.

The nature of the output motion of the actuator determines the mechanical drive train, so that would give our next defining factor.

Power Source	Size	Output motion		
		Multi-turn	**Quarter-turn**	**Linear**
Electric	Large	Worm/wheel	Worm/wheel + external QT box	Worm/wheel + ballscrew/acme
	Intermediate	Worm/wheel	Worm/wheel + external QT box	Worm/wheel + ballscrew/acme
		Spur gears	Spur gears	
	Small	Worm/wheel	Worm/wheel/ quadrant	Worm/wheel + ballscrew/acme
		Spur gears	Spur gears	Rack/pinion
		Epicyclic	Epicyclic	Integral motor/ ballscrew
Fluid Power	Large	Vane or lobe motor, worm/wheel	Scotch yoke	Piston
		Stepper	Rack/pinion	
			Gas/oil vane	
	Intermediate	Vane or lobe motor, worm/wheel	Scotch yoke	Piston
			Rack/pinion	
			Helical	
	Small		Scotch yoke	Piston
			Rack/pinion	Diaphragm
			Vane	

Figure 6.a. Matrix of actuator types.

The various types of mechanisms shown here are the more common types. There are actually many variations on these themes.

6.1   Electric

To simplify our understanding of the electric actuator, we can look at the two main parts of the drive train: the motor and the gearing mechanism. The output of an electric motor is used to drive the valve stem via a geared transmission mechanism. These gears take the high speed output of the motor and reduce the speed while increasing the torque. The output of the actuator then can engage and turn a stem nut or drive bushing to move the valve.

Torque and position sensors are used to monitor the output of the actuator. The torque sensors measure the amount of force delivered to the valve. The position sensors track the valve's position and its ends of travel.

The motor type used in an actuator is a function of the power supply as discussed in chapter 3. Large actuators, needing a greater power supply, have fewer choices of motor type when compared to smaller actuators.

6.1.1   Large electric

The larger sizes most commonly are driven by a 3-phase electric motor. These motors are robust and relatively trouble free. They can be designed to deliver the high torque needed for unsticking valves from their seats. When coupled directly to the worm shaft of a worm and wheel gear set, a simple and direct drive train is formed.

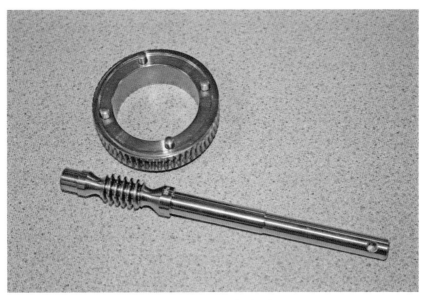

Figure 6.1.1.a. Worm and wheel gear set.
Image by permission of Rotork®.

Some designs incorporate spur or epicyclic gears between the motor and worm gear set to facilitate speed change or reduction.

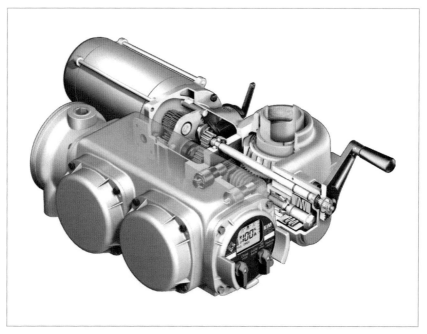

Figure 6.1.1.b. Electric multi-turn actuator with spur gears and output worm gear set. *Image courtesy of Emerson Bettis.*

6.1.1.1   Large electric multi-turn

The nominal speed of rotation of a two pole, 3-phase motor is 3,600 rpm, (for a 60 hertz power supply). To achieve a nominal output speed of say 60 rpm at the actuator output, the worm gear set would require a 60:1 reduction ratio. Changing the worm gear ratio would give differing output speeds and potentially change the final output torque.

The worm and wheel gear set is ideal for large actuator applications. Not only does it provide the needed reduction ratios and mechanical advantage, but it provides an irreversible self-locking drive (in the higher ratios).

The worm gear set has a low efficiency, usually in the range of 30% to 80% depending on the worm pitch. This has little relevance for most applications as an isolating actuator is infrequently operated so power consumed over time is low.

The gear sets on a valve actuator could be oil or grease lubricated, depending on the materials selected for the gears and the manufacturers preference.

The majority of these actuators, which are multi-turn design, are suitable for direct mount onto large and/or high-pressure gate and globe valves, sluice gates and guillotine dampers.

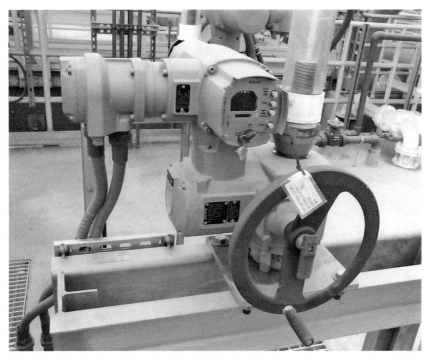

Figure 6.1.1.1.a. Multi-turn electric actuator on a sluice gate.
Image courtesy of Auma.

In some circumstances, a secondary bevel or spur gearbox is interposed between the multi-turn actuator and the valve. This is done for two possible reasons: to increase the torque from the primary multi-turn actuator or to increase the allowable stem acceptance so a large rising valve stem can be accommodated. It also may be the case that a combination actuator and secondary gearbox is more cost effective than a larger direct top-mounted actuator. This option is worth checking when sizing actuators for gate and globe valves in a cost-sensitive application or for an application that calls for a very slow speed of operation.

6.1.1.2　Large electric quarter-turn

To automate large quarter-turn valves such as pipeline ball valves and large butterfly valves, a secondary gearbox is often coupled to the multi-turn actuator to translate the multi-turn motion to a quarter-turn movement. This also increases the torque output to the valve by virtue of the mechanical advantage of the secondary gearbox.

Figure 6.1.1.2.a. Large electric multi-turn actuators with secondary quarter-turn gearbox on butterfly dampers. *Image courtesy of Auma.*

The secondary quarter-turn gearbox may have another advantage: it may provide a self-locking mechanism. If a quarter-turn valve is under pressure and has a natural tendency to move from this pressure or has a dynamic force acting on it from the flow of the pipeline media, then this could cause the valve to move. The self-locking mechanism of a valve actuator can hold the valve in position without the need for any power to be applied. This is important becasue the process can be assured of a stable valve position even in the event of power loss.

Figure 6.1.1.2.b. Large electric multi-turn actuator with secondary quarter-turn gearbox on a pipeline ball valve. *Image courtesy of Emerson Virgo.*

6.1.1.3   Large electric linear

In almost all cases with large electric actuators, the linear motion, when required by the valve's closure element, is provided by the stem nut of the actuator acting on the thread of the valve stem. This is a robust and reliable method of moving the valve. There are other benefits to the stem and nut mechanism. It adds mechanical advantage for generating thrust at the valve. It also provides a lock-in position tool by virtue of the irreversibility of the thread mechanism. The downside of this arrangement, is that the efficiency of the thread mechanism is low, and much of the actuator's power is lost to friction at the drive nut. This friction usually generates heat at the drive nut. In extreme cases, where long stems are used, this situation could accelerate nut wear. *A further explanation of the mechanics of drive nuts and valve stems can be found in appendix 14.1.*

6.1.2   Intermediate electric

6.1.2.1   Intermediate electric multi-turn

Smaller versions of the large electric worm and wheel actuator are available, usually within the same product range as the large electric units. These could be powered, not only by 3-phase motors, but also 1-phase AC or even DC motors due to the lower power demanded by the valve.

6.1.2.2   Intermediate electric quarter-turn

The practice of adding a secondary gearbox for quarter-turn valves is also common on intermediate sizes. However, as the size reduces, many actuator ranges transition into a different drive train configuration. Some designs add a second integral worm and quadrant to produce the quarter-turn actuator variant.

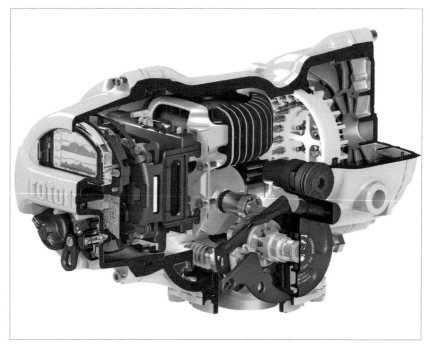

Figure 6.1.2.2.a. Intermediate size electric quarter-turn actuators.
Image by permission of Rotork®.

Figure 6.1.2.2.b. Intermediate size electric quarter-turn actuator.
Image courtesy of Flowserve limitorque.

An alternative is to use an all spur gear drive train with a spur quadrant to give a final quarter-turn output motion.

6.1.2.3 Intermediate electric linear

For linear drive requirements at the intermediate sizes, a multi-turn actuator could have a linear drive mechanism attached to its base. This could take the form of a ball screw drive, which is efficient but reversible, or the less efficient but irreversible threaded stem, similar to the stem of a valve.

Figure 6.1.2.3.a. Electric linear valve actuator showing the internal control circuitry. *Image courtesy of Harold Beck & Sons Inc.*

Figure 6.1.2.3.b. Electric linear valve actuator mounted on a sliding stem control valve. *Image courtesy of Harold Beck & Sons Inc.*

6.1.3 Small electric

Very small electric actuators are often integral in design with the valve. An example is some solenoid-operated valves. The actuators in these cases are simple solenoids.

The largest variety of drive configurations is found in the small electric category.

6.1.3.1 Small electric multi-turn

The main drive train configurations tend to use basic spur gears, but epicyclic, hypocycloidal, nutating and other gear combinations are also used.

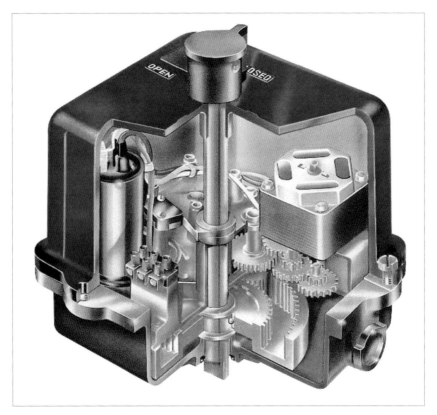

Figure 6.1.3.1.a. Small multi-turn actuator with spur and epicyclic gear drive train. *Image courtesy of Flowserve Limitorque.*

6.1.3.2   Small electric quarter-turn

The worm and wheel type drive train is often used for this classification. However, some manufacturers utilize their multi-turn range, restricted to a quarter-turn output.

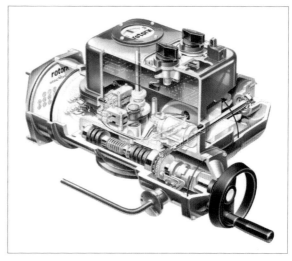

Figure 6.1.3.2.a. Small electric unit with primary worm gear reduction and worm quadrant output.
Image by permission of Rotork®.

6.1.3.3   Small electric linear

For linear outputs, as well as the linear drives seen in the intermediate sizes, there are rack and pinion drives with spur gears providing the preliminary reduction.

Other designs use a ball screw for the final linear output. Also, the threaded shaft and drive nut mechanism is used in several small linear actuator designs.

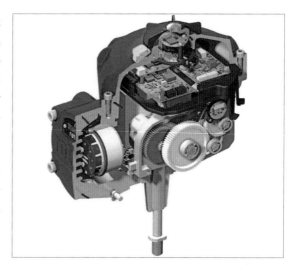

Figure 6.1.3.3.a. Small electric linear output actuator using a rack and pinion final drive.
Image by permission of Rotork®.

6.2   Fluid-powered

Similar to electric actuators, the fluid power actuators can be divided into two main components, the fluid power chamber and the motion mechanism.

The power chamber for linear actuators could be a piston/cylinder assembly or a diaphragm. For part-turn actuators the power chamber could be one of the linear types coupled to a rotary conversion mechanism or a vane mechanism. For multi-turn devices a lobe, gear or vane motor might be used. In almost all cases the power could be provided by hydraulic fluid or instrument air.

The piston and cylinder is used in many large and small actuators. The fluid pressure acting on the surface area of a piston generates a force in direct proportion to the magnitude of the pressure and the projected area of the piston.

> Force = pressure x projected area.

So, an instrument air supply of 80 psi acting on a 6-inch diameter piston will generate a force of $80\pi d_2/4$ or 2,260 lbs. This force could act directly on a sliding valve stem or be used to drive through a scotch yoke or other mechanism to rotate a part-turn valve stem.

6.2.1   Large fluid powered

6.2.1.1   Large fluid powered multi turn

Some pipeline gate valves are too large to be automated by a top mounted linear cylinder actuator. The cumulative height of the valve plus cylinder may be structurally unstable or may interfere with other structures. In these circumstances there are alternative fluid-powered actuator types that have a multi-turn output.

The first is a fluid-powered version of the electric-driven worm and wheel multi-turn actuator described in section 6.1.1. Typically the pneumatic or hydraulic motor is substituted for the electric motor. The fluid-powered motor can be controlled by fluid or electric switches. If an instrument electrical supply is available, then electrical torque and limit switches can be used to control a solenoid-activated fluid power direction control valve.

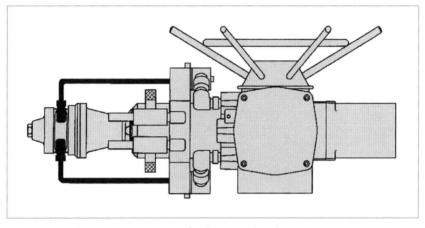

Figure 6.2.1.1.a. Large fluid-powered multi-turn actuator.
Image by permission of Rotork®.

The stepping type of fluid-powered multi-turn actuator uses a ratchet mechanism to rotate a wheel connected to the valve stem nut. There are one or more pistons that drive the mechanism on the periphery of the driven wheel.

Although more compact than the linear cylinder actuator, these devices drive valve movement through a stem nut. Because the stem nut is far less efficient than the direct drive of a linear actuator, these types of stepping actuator mechanisms have a much higher power consumption and likely provide a longer operating time to stroke the valve.

Figure 6.2.1.1.b. Large fluid-powered stepping type multi-turn actuator on a gate valve. *Image by permission of Rotork®.*

6.2.1.2   Large fluid powered quarter turn

To provide a quarter-turn motion for large ball, butterfly and plug valves as well as quarter-turn louver and butterfly dampers, a quarter-turn mechanism is coupled to a linear cylinder. The most popular of these mechanisms is the modified scotch yoke. (The original scotch yoke converts rotary motion to linear).

Scotch yoke actuators can be configured as double-acting or spring return. Some of these machines are among the largest actuators made.

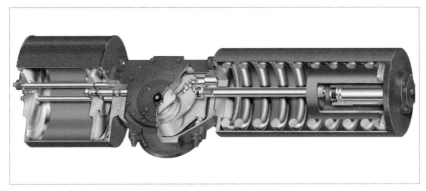

Figure 6.2.1.2.a. Spring return scotch yoke actuator.
Image courtesy of Emerson Bettis.

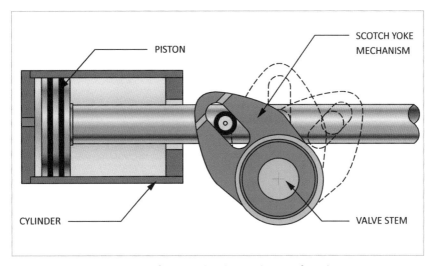

Figure 6.2.1.2.b. Modified scotch yoke mechanism for valve actuation.

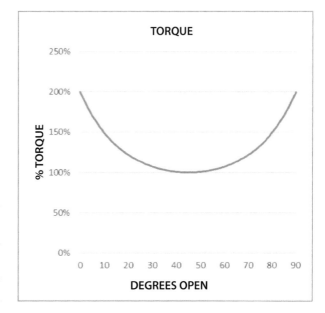

Figure 6.2.1.2.c.
Modified scotch
yoke mechanism,
double-acting
actuator torque
characteristic.

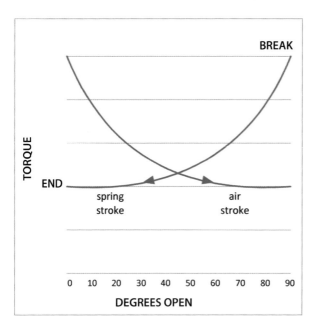

Figure 6.2.1.2.d.
Modified scotch yoke
mechanism,
spring return
actuator torque
characteristic.

At the beginning of the stroke, the torque is amplified by virtue of the geometry of the scotch yoke mechanism. This gives the needed boost of torque to move the valve off its seat, the point of most friction. Further modification to the geometry of the scotch yoke by "canting" the slot can provide even more of a characteristic boost.

For actuators with the same cylinder diameter and body size, an actuator built as a spring return type will have an end of stroke torque which is generally one third of the equivalent double-acting type actuator end torque. On a double-acting actuator, all of the piston force is transmitted to the scotch yoke mechanism to move the valve. On a spring return actuator of the same size, a large part of the piston force is used to compress the opposing spring during the spring stroke, and the remaining force is available to move the valve.

For larger valves with a really high torque demand, a combination of two mirror image scotch yoke mechanisms in a quad configuration can more than double the torque output. The balanced mechanism reduces the friction torque on the scotch yoke output bearings, thereby providing more than twice the output force achieved by just doubling the cylinders.

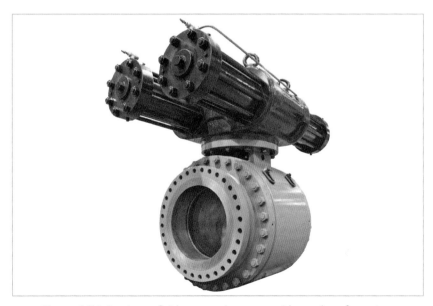

Figure 6.2.1.2.e. Large fluid-powered actuator with quad configuration.
Image by permission of Rotork®.

Figure 6.2.1.2.f. Large spring return scotch yoke pneumatic actuator.
Image by permission of Rotork®.

There are alternatives to the scotch yoke mechanism for translating the linear motion of a hydraulic or pneumatic piston. The rack and pinion mechanism is occasionally used on large valves. Unlike the scotch yoke mechanism, the output force of the rack and pinion is constant throughout its stroke. For subsea valves the high-pressure hydraulic rack and pinion design is often selected because of its robust and compact mechanism and body.

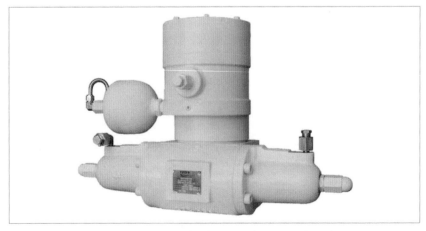

Fig 6.2.1.2.g. Rack and pinion fluid-powered actuator for subsea service.
Image by permission of Rotork®.

Large vane actuators also deliver high torques. These machines utilize a double vane configuration in a vertical cylinder but unlike the conventional piston/cylinder actuator, the vane actuator has a coaxial rotating (quarter-turn) output shaft. Using high-pressures they can operate large pipeline valves. To enhance the sealing around the vanes at these high-pressures, hydraulic oil is used in the actuator cylinder.

Figure 6.2.1.2.h. Large vane-type fluid-powered actuator.
Image courtesy of Emerson Bettis Shafer.

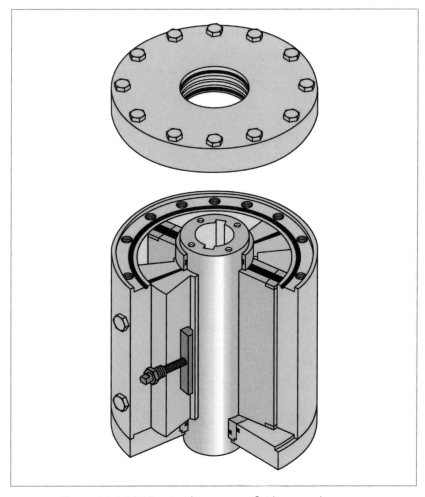

Figure 6.2.1.2.i. View inside vane-type fluid-powered actuator.
Image courtesy of Emerson Bettis Shafer.

When operating from the medium in a gas pipeline, both vane and scotch yoke actuators often use intermediate gas/oil tanks to transmit the gas line pressure to the actuator. They are commonly known as "gas over oil" actuators. The oil not only lubricates and reduces any leakage across the seals, but also protects the internal moving parts from any corrosive action by the pipeline gas. The gas could contain moisture and traces of hydrogen sulfide, which could accelerate corrosion.

Figure 6.2.1.2.j. Large gas-over-oil scotch yoke quarter-turn actuator.
Image by permission of Rotork®.

6.2.1.3   Large fluid powered linear

The most common prime mover of large linear fluid power actuators is the basic piston and cylinder. This configuration is not only mechanically simple but also very efficient. Fluid power is directly translated into linear mechanical movement at the valve stem. The high efficiency, however, means that the actuator is not inherently self-locking. This, if needed, would have to be provided by the fluid control circuitry.

There are two basic configurations for linear cylinders:

1) Double-acting, where the fluid power is used for stroking in both directions
2) Spring return, where an opposing spring is used for stroking in one direction and fluid power for the return direction

Figure 6.2.1.3.a. Large fluid-powered double-acting linear actuator on a gate valve. *Image by permission of Rotork®.*

6.2.2 Intermediate fluid powered

Large fluid power devices are often scaled up versions of the intermediate category of fluid power actuators. The intermediate size actuators use a reduced piston area, or they have applied pressure that is less compared to the large category. Reducing pressure allows a thinner-walled cylinder to be used. Reducing the piston size correspondingly reduces the envelope dimensions. Both have the effect of reducing actuator weight and cost.

Figure 6.2.2.a. Pneumatic spring return intermediate-size actuator.
Image by permission of Rotork®.

This intermediate category introduces an interesting mechanism, the helical hydraulic actuator. This device delivers a quarter- or part-turn output from the action of a piston on a helix mechanism. These actuators are compact and are often used in marine applications.

The smaller fluid power actuators similarly scale up to reach into the intermediate-size area.

6.2.3   Small fluid powered

6.2.3.1   Small fluid powered linear

Perhaps the most ubiquitous linear pneumatic actuator is the spring diaphragm valve actuator. This device is often an integral part of a process control valve assembly supplied by the control valve maker.

A flexible composite diaphragm is enclosed in a pressed steel body and attached to an output shaft. Pressure applied either under or on top of the diaphragm moves the output shaft up or down. An opposing spring returns the actuator to its original position.

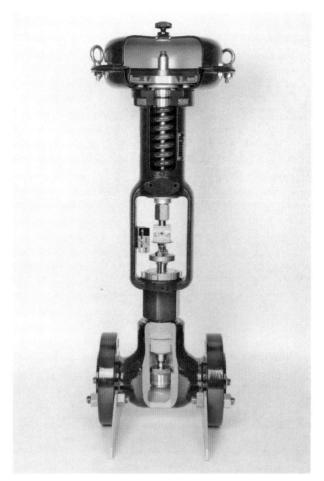

Figure 6.2.3.1.a. Linear pneumatic spring diaphragm actuator on a plug valve (air to open, spring close). *Image courtesy of Richards Industries.*

This design lends itself well to the modulating control required by process control valves. There are few moving parts and little friction so the hysteresis of the assembly is low. This is important for precise positioning of control valves. Originally, they were designed to operate via a 3 to 15 psi control air signal that was applied directly to the diaphragm. To reduce the size of the diaphragm on the actuators, a positioner was interposed on the actuator between the control air signal of 3 to 15 psi and the instrument air power supply at about 80 psi. The higher pressure instrument air delivers more thrust for a given diaphragm size.

6.2.3.2   Small fluid powered quarter turn

Linear spring diaphragm actuators have also been coupled to rotary drive mechanisms to provide a quarter-turn output.

Figure 6.2.3.2.a. Spring diaphragm quarter-turn pneumatic actuator on a segmented ball valve. *Image courtesy of Richards Industries.*

Another design of the smaller quarter-turn fluid-powered actuators commonly utilizes aluminum extrusions to provide a one-piece body and cylinder assembly. Mostly these are used for isolating duty, but it is becoming more common to see them used for modulating applications.

These designs are inexpensive to manufacture. Two pistons with integrated rack gearing are mounted in opposition and provide a balanced force on the central pinion output shaft, which is connected to the valve stem.

Spring return action is provided by a set of springs on the outboard side of the opposing pistons.

The torque characteristic of the rack mechanisms provides a flat or constant torque output in the double-acting configuration.

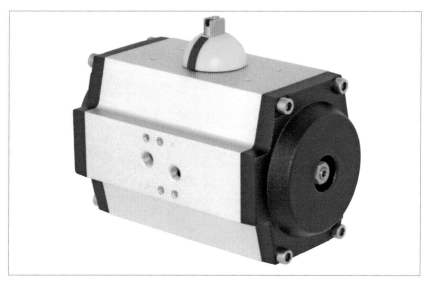

Figure 6.2.3.2.b. Rack and pinion double-acting pneumatic actuator.
Image by permission of Rotork®.

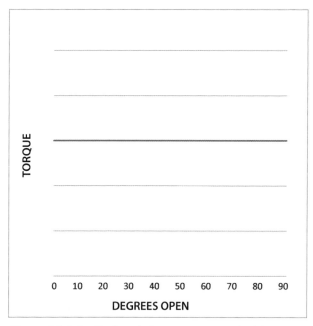

Figure 6.2.3.2.c. Rack and pinion pneumatic double-acting
actuator flat torque characteristic.

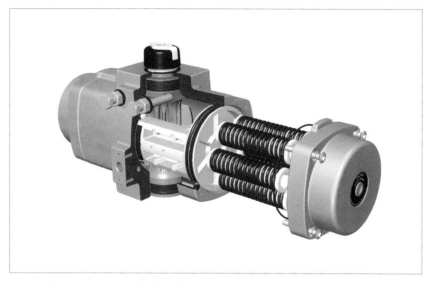

Figure 6.2.3.2.d. Rack and pinion pneumatic actuator spring return.
Image courtesy of Emerson Bettis.

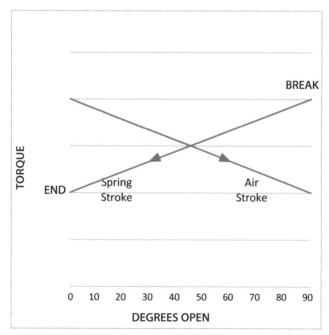

Figure 6.2.3.2.e. Rack and pinion pneumatic spring return actuator torque
characteristic.

On the uni-body design the forces on the pinion are balanced axially as the pistons are opposed. However, the offset of the center of piston force to the rack and pinion contact point generates a moment with a resulting side force. This is often restrained in these designs by a nylon pad rubbing on the cylinder wall. It also compounds with the separating force between the rack and pinion gears to increase the friction forces within the actuator.

Some designs of the rack and pinion actuator axially align the center of pressure of the piston with the contact point of the rack and pinion gear, eliminates all side forces with the exception of the rack and pinion separating force. There is a resulting improvement in hysteresis and positioning accuracy. This is not a uni-body type construction, but usually a cast body construction.

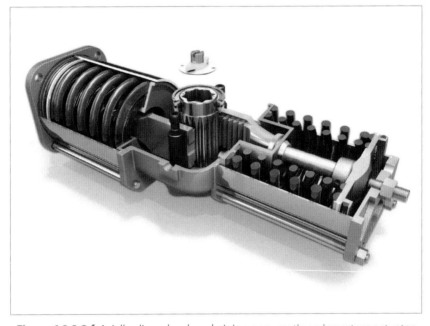

Figure 6.2.3.2.f. Axially aligned rack and pinion pneumatic spring return actuator. *Image courtesy of QTRCO.*

The uni-body design is not limited to the rack and pinion mechanism: there are also balanced scotch yoke designs that utilize these types of body extrusions. These designs, however, give the actuator a typical scotch yoke torque output.

This can prove useful for some valves with high break-open torque demand.

Generally, with smaller quarter-turn valves (below about 2″), the torque demand tends to flatten out over the valve stroke. This is because the effects of friction on the stem and seats throughout the stroke become proportionately higher when compared to the break-open torque of the valve.

Figure 6.2.3.2.h. Small vane pneumatic actuators.
Image by permission of Rotork®.

The small vane actuator design is very different from for large pipeline vane designs. These small vane designs are usually pneumatically powered. The vane is a single paddle sweeping a quarter-turn arc in an aluminum chamber. The spring return variant uses a clock spring-type mechanism coaxially mounted on top of the vane body to provide the return stroke.

The torque characteristic of this design of double acting vane is flat, just like the small rack and pinion torque characteristic in figure 6.2.3.2.c. Similarly the spring return torque characteristic is like figure 6.2.3.2.e.

There are other novel designs for small pneumatic actuators using cams as an alternative to the rack mechanism. Similarly, some designs use a bladder to move a lever arm through the quarter-turn arc.

6.3   Hybrid actuators, electro-hydraulic

While briefly described in previous sections, the combination of a fluid power actuator driven by a self-contained electro-hydraulic power unit is a viable solution for valve automation requirements. A pipeline shutdown application is one application where this applies. The electro-hydraulic actuator also can provide quick stroke times and precise positioning for some specialized applications such as boiler feed pump valves, fuel valves and anti-surge valve applications.

Figure 6.3.a. Electro-hydraulic actuator on a boiler feedwater recirculation valve.
Image courtesy of REXA Inc.

One advantage of this type of actuator is the ability to select a DC or AC motor to suit the available power supply or to condition the available supply to suit the motor. Since these actuators are self-contained, they can be powered the same way as an electric actuator, but they also have the ability to automate a valve in the save way as a double-acting or spring return hydraulic actuator.

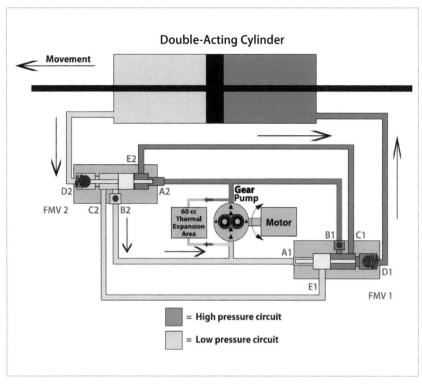

Figure 6.3.b. An electro-hydraulic actuator hydraulic circuit schematic.
Image courtesy of REXA Inc.

As illustrated above, to move the cylinder piston to the left, the pump/motor turns in the direction to pressurize FMV-2 (flow matching valve) through port A2. The spool in FMV-2 becomes unbalanced by the pressure differential and moves to the left, lifting its check valve, opening port D2 to port B2 and port A2 to port E2. High-pressure fluid flows through Port E2 to the right side chamber of the cylinder. Since the hydraulic circuit is closed, the same amount of oil that flows into the right side of the piston must be extracted from the left side. This allows oil movement without an active reservoir. This oil flows through the open check valve of FMV-2 and into pump suction.

Other types of electro-hydraulic actuators use an oil reservoir (to compensate for the piston rod volume) and solenoid-operated direction control valves similar to a conventional hydraulic actuator.

For oil and gas transportation applications, self-contained large electro-hydraulic actuators can provide spring-powered shutdown capability for pump station valves and sectioning valves. The spring can move the valve to the required position during an emergency, in the event of a loss of main power supply.

Figure 6.3.c. Large electro-hydraulic actuator on a pump station ball valve. *Image by permission of Rotork®.*

Chapter 7

Fail to position

— ◆ —

One of the options required of an actuator is the ability to go to a preset position in the event of a plant-defined failure event. For purposes of actuator selection, there are only two types of failure definitions that we are concerned with.

A)   A failure that triggers a "go-to-fail position" control signal to the actuator.
B)   A failure of the power supply to the actuator itself.

The type (A) of fail-to-position signal is relatively easy for any actuator to accommodate. It simply sees a command from the control system to go to a specific position and moves the valve to that position using the available normal power supply.

Alternatively, there may be a loss of command signal condition. In this case the actuator controls can be configured to send the actuator to an open, close, or intermediate position or even to stay put. This requirement should be specified at the time of procurement. Although relatively easy to accommodate, some configuration of controls would be required, particularly for fluid-powered actuators.

Another type of fail signal is an emergency shut down (ESD).This type of signal often has its own dedicated wires from the control room, or a separate

ESD control system to the actuator. The ESD signal tells the actuator to go to a preset position regardless of any other control signal coming from the other control wiring. Pneumatic actuator controls can have a dedicated solenoid valve for ESD that vents air from the actuator to provide the fail-to-position mode. Some electric actuators can provide dedicated terminals that activate a preprogrammed ESD function in the actuator.

The type (B) failure mode, fail to position on loss of actuator power supply, is more costly to accommodate.

On large 3-phase electric actuators, it is not usually practical to store enough energy to drive a valve to a fail close or open position. This is because the combination of actuator and valve stem drive train has such a low efficiency. Conversely, the beneficial aspect of this actuator type is its ability to fail in last position and maintain that position because of the irreversibility of the drive train.

Intermediate and small electric actuators have several possible solutions.

Some actuators utilize a clock spring-type mechanical energy storage device to drive a conventional electric spur-geared actuator. There are also a quarter-turn fail safe actuators that use the torsional mode of a coil spring to store energy. These devices, however, do not scale up to the larger valves very well.

Electrical energy storage in the form of lead acid accumulators or batteries, has been in use for DC motor actuators for some time. Where space is not an issue, this is an effective but expensive solution. Some power stations, for example, have a "black start" capability using a room full of lead acid accumulators to drive DC motors for starting up the plant, should the main power grid be lost.

For small actuators an alternative electrical energy storage method is to use super capacitors. These are compact devices that can deliver energy rapidly to drive a DC motor. They are frequently used on electric control valves to replicate the fail-to-position capability of the conventional spring diaphragm actuator.

For fluid-powered actuators, particularly quarter-turn and diaphragm actuators, the fail to position on loss of power is far easier to achieve. .

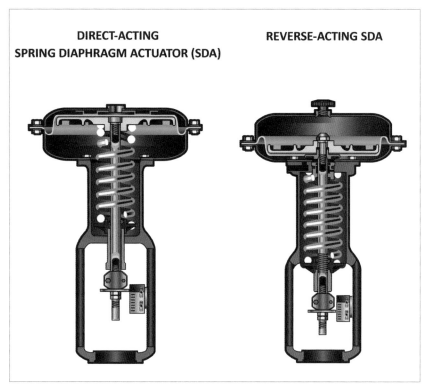

Figure 7. Pneumatic spring diaphragm actuators.
Direct acting, air close, spring open. Reverse acting, air open, spring close.

A popular solution is to oppose the fluid power cylinder with a spring to provide a spring return action. The actuator is built at the factory to be either fail open or fail close depending on the application. The simplicity of these devices makes them reliable and dependable. Frequent use in the oil and gas industries, where safety shut-down is engineered into plants, has made them the default actuator for process shutdown and ESD valves, particularly for offshore applications.

Where springs are not suitable, either because of space constraints or because multiple valve strokes are required, then fluid power accumulators can be used. For pneumatic actuators an air receiver tank is frequently used; for hydraulic actuators a hydraulic accumulator with gas pressurization is used with a gas pre-charged bladder device. An alternative accumulator design uses a gas pre-charged cylinder and floating piston as a separator between the gas and the hydraulic oil.

Calculating the size of the tanks can be achieved using formulas that consider the swept volume of the actuator cylinder, the required pressure to stroke the valve and the available supply pressure. *See appendix 14.8 for details.*

Chapter 8

Actuator selection, the SIMPLE method

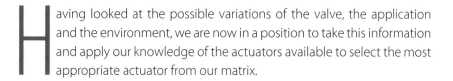

Having looked at the possible variations of the valve, the application and the environment, we are now in a position to take this information and apply our knowledge of the actuators available to select the most appropriate actuator from our matrix.

The three main selection parameters are detailed below, but the table gives us an overview and guide to the options available.

Selection 3	Selection 1	Selection 2		
Power Source	**Size**	**Output motion**		
		Multi-turn	**Quarter-turn**	**Linear**
Electric	Large	Worm/wheel	Worm/wheel + external QT box	Worm/wheel + ballscrew/acme
	Intermediate	Worm/wheel	Worm/wheel + external QT box	Worm/wheel + ballscrew/acme
		Spur gears	Spur gears	
	Small	Worm/wheel	Worm/wheel/ quadrant	Worm/wheel + ballscrew/acme
		Spur gears	Spur gears	rack/pinion
		Epicyclic	Epicyclic	Integral motor/ ballscrew

Selection 3	Selection 1	Selection 2		
Power Source	**Size**	**Output motion**		
		Multi-turn	**Quarter-turn**	**Linear**
Fluid Power	Large	Vane or lobe motor, worm/wheel	Scotch yoke	Piston
		Stepper	Rack/pinion	
			Gas/oil vane	
	Intermediate	Vane or lobe motor, worm/wheel	Scotch yoke	Piston
			Rack/pinion	
			Helical	
	Small		Scotch yoke	Piston
			Rack/pinion	Diaphragm
			Vane	

Figure 8. Table of actuator type and drive mechanism.

8.1 SIze – Selection 1

The valve size and other attributes dictate the amount of force required to move it through its stroke. The table above gives a rough indication of the types of actuator available. Once we know the value of the force needed, either from the valve maker, our reference data or calculation, we can then start the specific actuator size selection process.

Most actuator manufacturers have tables of torque or force data for actuator size selection. Both fluid power and electric actuator data can be reviewed. (If the required power source is already a given, this knowledge would reduce the possible variables significantly by narrowing the options for selection of parameter 3).

8.2 Motion – Selection 2

The required motion of the valve input will narrow down the list of suitable actuators to linear, quarter-turn and multi-turn. As can be seen from Figure 8, this leads us to the mechanical arrangement of the actuator drive.

8.3   Power – Selection 3

Often the power supply choices are limited. There may be instances where there is more flexibility so we should review the choices next and make a selection or either fluid power or electric power.

In circumstances where only a low-wattage power supply is available, then the efficiency of the actuator may factor into the selection. A more efficient drive train in an actuator can significantly reduce the power demanded to cycle the valve. For example, a fluid-powered shutdown scotch yoke actuator is far more efficient than an air motor-driven multi-turn worm and wheel drive actuator. Similarly, for electric actuators a spur-geared quarter-turn unit is generally more efficient than a worm and wheel unit. The following table provides a guide to the approximate efficiencies of each type of actuator mechanism, although this varies with manufacturer. For applications where electric power consumption is of concern, it is best to consider the current draw data provided by the manufacturer.

Eficiencies of various actuator mechanisms			
Mechanism	Low	High	median
Worm and wheel	50%	90%	70%
Spur gear	90%	95%	93%
Bevel gear	89%	94%	92%
Electric motor	70%	90%	80%
Fluid power rotary motor	60%	80%	70%
Acme thread and nut, irreversible	15%	35%	25%
Scotch yoke mechanism	50%	75%	63%
Vane actuator	70%	80%	75%
Instrument air system	50%	70%	60%

Figure 8.3. Actuator mechanism efficiencies.

8.4   Location and Environment

Not all actuators are suitable for hostile or explosive environments. Similarly, not all environments require the expense of actuators designed with exotic corrosion-resistant or specially prepared coatings. The level of environmental protection is described by the NEMA or IP ratings we saw in section 4.

This last selection is both a technical and commercial decision. Usually the lower the protection rating, the less expensive the actuator. However it would be a false economy to underrate the environmental protection. Moisture can quickly disable electrical circuits and switches if they are not protected by a suitable enclosure.

See appendix 14.1 for sizing guides and appendix 14.8 for sample actuator specification and data sheets.

Chapter 9

Digital communications

— ◆ —

I n section 5.4.3 we looked at how external control systems have improved from the 3 to 15 psi control signal of the last century, to the digital communications systems of today.

In the actuator world, both electric- and fluid-powered actuators can be controlled and monitored with the digital field bus systems commonly used for other field instrument devices. The more commonly used systems for actuators are:

1) HART
2) Foundation Fieldbus
3) Modbus
4) Profibus
5) Devicenet
6) Actuator manufacturer's proprietary systems

Actuator manufacturers usually have a suite of optional communications boards or chips that can be incorporated into the build of the actuator during manufacture or sometimes can be retrofitted later. These interface boards or chips typically allow a standard actuator to be monitored and controlled via an open architecture communications system or the manufacturer's own proprietary system.

Open architecture systems allow any manufacturer to provide equipment that can communicate over the common network. This means that on any given network there may be equipment from multiple manufacturers communicating together.

Proprietary systems usually only support the equipment of a particular manufacturer.

For valve actuator applications, it is important that a robust communications system is used. An important aspect is the system's tolerance to electromagnetic interference from nearby equipment such as motors for pumps or fans. Some systems have built-in redundancy, so if there is a network fault or a communications card fails, communication is maintained to the connected devices.

Almost all of the digital communications systems used by valve actuators carry the signal over hard-wired connections. There are some wireless systems (not to be confused with blue tooth setup communications) that employ mesh network technology such as "ZigBee." But these are rare. At this point there seems to be a general reluctance on the part of end users to entrust valve control to purely wireless communications.

A brief description of the various open systems follows:

HART is the acronym for Highway Addressable Remote Transducer. This is perhaps the most frequently used system for process control instrumentation. For actuators used in process control applications, a HART interface capability is important.

The HART communication is transmitted as a superimposed digital signal over a standard 4 – 20 mA control signal. Because many control systems used this 4 – 20 mA signal as a basic standard for analog control, it was an easy step for process plants to move from analog to this form of digital communications. Many plants have installed smart positioners using an analog 4 – 20 mA control signal that are also HART capable in anticipation of moving to full digital communications in the future.

<u>Foundation Fieldbus</u> is an entirely digital communications system. The field implementation for devices is designated H1 and communication is via a shielded twisted pair of wires. Actuators interface to the twisted pair via an H1 interface card.

Figure 9.a. Fieldbus digital valve positioning unit on sliding stem control valve.
Image courtesy of Richards Industries.

<u>Profibus</u> originated in Germany and has been supported by Siemens. It has arguably the largest installed base of industrial network nodes in the world. There are two main field level nodes, Profibus, PA and Profibus DP.

PA (process automation) can be used as an intrinsically safe communication and power system for sensing and measurement instruments.

DP (decentralized peripherals) uses the same protocol as PA, but is more suited for actuators, motors and sensors.

Modbus was originated by Modicon and is an easily understood general purpose industrial network protocol. It can be used to communicate directly with valve actuators, but is less frequently selected over other field communications systems.

DeviceNet was developed by Allen-Bradley and is now an open protocol supported by the ODVA (Formerly Open DeviceNet Vendors Association). Like Modbus, it may be used for actuator communications, but it is not as commonly used as some others.

Chapter 10

The interface between valves and actuators

— ◆ —

Valve adaption hardware is the mechanical interface between the actuator and the valve. It consists of two main parts: a static component and a dynamic component.

10.1 Static mounting

Static mounting is the fixed connection between the valve top-works and the actuator base. For many actuators there are applicable standard dimensions for the static base fastening. These are detailed in the European metric standard ISO 5211 for part-turn valves, ISO 5210 for multi-turn valves and the U.S. standard MS SP-101 in imperial units.

These standards relate the flange mounting dimensions, such as bolt size and pitch circle diameter, to the maximum allowable transmitted force.

The objective is to match the valve flange to the actuator flange so the two can be simply bolted together.

Figure 10.1.a. Electric actuator with direct mounting to the flange of a gate valve. *Image by permission of Rotork®.*

In many circumstances this can be achieved, but sometimes an additional adaption flange or bracket is needed if the valve top-works are not to the standards or there is a need to move the actuator away from the valve. This could be because of the presence of packing on the top-works or physical interference with part of the actuator. Other reasons include, moving the actuator away from the high temperature of the valve or for galvanic insulation. (*See appendix 14.6*)

Figure 10.1.b. Pneumatic rack and pinion actuator on a ball valve with box type adaption bracket. *Image by permission of Rotork®.*

Figure 10.1.c. Box bracket with stem coupling and mounting hardware. *Image by permission of Rotork®.*

10.2 Dynamic stem coupling

The second part of the valve adaption is the moving coupling. This part is the dynamic coupling between the valve stem and the actuator output.

As discussed in chapter 2, multi-turn electric actuators usually incorporate a drive bush. This drive bush must hold the valve stem in position by absorbing the thrust reaction from the valve closure element. To do this the drive bush is restrained by bi-directional thrust bearings in the actuator assembly. Some older designs had taper roller thrust bearings at the top and bottom of the actuator gear case. More recent designs contain the thrust in the base of the gear case. The more convenient thrust containment method is to have a detachable thrust base as shown in Figure 10.1.a. This has the advantage of allowing the actuator to be removed and leaving the thrust base in place on the valve to hold the stem in position. It also allows multi-turn actuators to be mounted on secondary gearboxes without the thrust base because this is not needed. This saves cost and weight on the assembly.

On part-turn valves the dynamic coupling takes the form of a machined bushing. The valve stem will determine the type of machining needed on the coupling. The stem configuration is often a keyed shaft with a standard key or woodruff key. On smaller valves the stem could be machined with a square top or opposing flats.

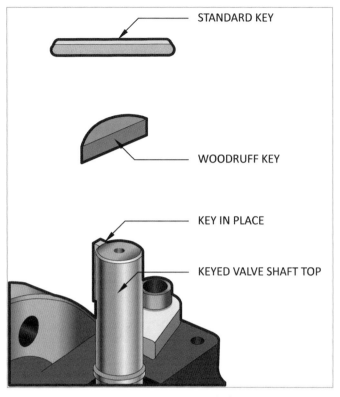

STANDARD KEY

WOODRUFF KEY

KEY IN PLACE

KEYED VALVE SHAFT TOP

Figure 10.2.a. Keyed shaft.

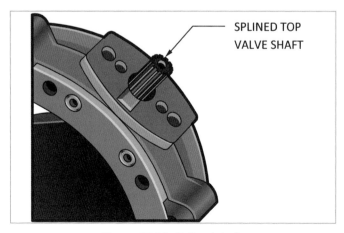

SPLINED TOP
VALVE SHAFT

Figure 10.2.b. Splined shaft.

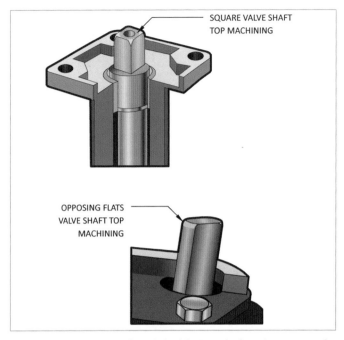

Figure 10.2.c. Square shaft and double D or shaft with opposing flats.

For multi-turn valves the actuator drive nut is usually machined from aluminum bronze for rising threaded stem valves. This provides good mechanical strength and wear resistance for the relative movement between the threaded stem and the bush.

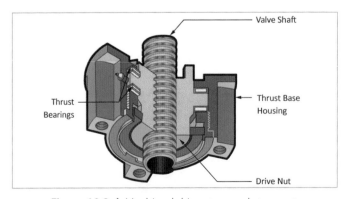

Figure 10.2.d. Machined drive stem and stem nut.

Chapter 11

Retrofit

— ◆ —

11.1 What is retrofit?

Figure 11.1.a. Mounting a fluid-powered actuator on a plug valve.
Image by permission of Rotork®.

Valve actuator retrofit is the practice of selecting and installing a valve actuator on an existing valve. Often the valve may already have a manual drive or an inoperable or obsolete actuator in place.

Existing installations can have their lives extended by retrofitting new valve actuators and controls.

Figure 11.1.b. New electric actuator being retrofitted to a valve in an existing plant. *Image by permission of Rotork®.*

Many facilities have valves capable of being automated and can benefit from that automation. For example, an older plant with dozens of valves could upgrade an antiquated control system to new technology with powerful electronics, advanced diagnostics capabilities and new actuators on the existing valves.

Another example would be the retrofit of a single, manually operated valve that takes a lot of time and effort to open and close. Such a valve would benefit from the faster and easier operation that would come from an actuator.

11.2 The benefits of retrofit

Benefits from retrofits accrue over time as the plant operates more efficiently. Specifically, when installing retrofit equipment, plant owners replace manually operated valves or old technology with modern fluid-powered or electric-powered valve control equipment.

A cost benefit results from upgrading an existing plant with valve automation. What is more, it may not be necessary for older plants to engage in a major pipework hardware renovation or reengineering of existing processes to upgrade to automated valve operation.

Integration of electric or fluid power actuators with digital control systems results in a control automation upgrade that could give a cost-effective life extension to existing facilities.

Some valves take a long time or much effort to operate manually. These are good candidates for single retrofits using the applied power of a valve actuator. The speed of operation of any particular valve can be improved without the constraint of the space needed for workers to gather round and operate a handwheel.

Retrofitted valves can be controlled by a central system. This facilitates the maximum output of a plant or process. Once a plant or a section of a plant is retrofitted with valve actuators and centralized control is introduced, the throughput of the plant increases and work hours spent manually operating valves decreases or are eliminated. Retrofitted and automated plants frequently can run for an increased number of hours each day with fewer operators, further increasing return on investment. By coordinating and effectively controlling the process, production is optimized and waste reduced.

The advances made in valve actuator technology can provide more reliable process operation. Heavy-duty, industrial valve actuators are now capable of prolonged and reliable service in hostile environments and hard-to-reach locations.

Control and supervisory systems can collect data from valve actuators to allow continuous and comprehensive monitoring of the automated valve operational status. This data can be integrated into an asset management system to allow planning of preventive and predictive maintenance.

Frequently, valves are located in extremes of temperature or in toxic environments that are hazardous to personnel. Through retrofits, valves in hostile environments can be operated easily without placing personnel at risk or discomfort.

At many plants, valves may be located high up in the structure on boilers or vessels such that some time may be needed to reach them. They may also be located in sumps or pits that are also difficult to reach. Retrofitted valves can be controlled from a distance, eliminating the need for a local presence. By automating the operation of these valves, personnel time and effort can be better utilized.

Manually operated valves often depend on the subjective assessment of the valve position by the operator at the valve turning the handwheel. This information might be conveyed verbally to a control room and may be needed for safe operation of a process. However, the positon assessment is subject to human error, and the actual valve may be mistakenly reported as another, adjacent valve. Retrofitting a valve actuator provides unambiguous feedback of the position of the valve using the position switches of the actuator. This data is readily available both locally and remotely to facilitate the process and for safety interlocks.

When valves are automated then instantaneous action can be taken to contain the impact of upsets or emergencies. For example, critical valves can be driven by their actuators to a preferred safe position. Automated equipment also can react rapidly to open or close valves under emergency situations, thereby avoiding spills and other accidents. Process controllers can be used to analyze the status of flow control elements and make the correct selection of which valve to actuate to mitigate accidents.

11.3 Retrofit procedures

Selecting actuators for retrofitting is similar to the "SIMPLE" selection process, but with the added challenges of determining the sizing requirements and the hardware needed to mount the actuator onto the existing valve.

One challenge is that there may be no information available on the force demand of the valve. That may be because the valve manufacturer is out of business or unable to provide the information required. If so, then an educated assessment of the valve force demand must be made.

The physical mounting of the actuator to the valve usually requires the design of a mounting bracket for static mounting together with a dynamic coupling to the valve stem for movement of the valve closure element.

Control of the actuator will require integration into the existing or upgraded control architecture. Although most actuators have local controls available, some form of remote control is usually employed.

11.3.1 Retrofit actuator selection

Actuator selection for retrofit purposes can follow the SIMPLE method described in chapter 8. If the force required by the valve is not available from the valve manufacturer, then the force can be estimated for gate and globe valves by using the sizing calculation shown in appendix 14.1. The thrust needed to close the valve against the pressure in the pipeline is calculated first. This will be added to an estimate of the stem packing friction.

To convert the linear valve force demand to a rotary torque demand value, the "stem factor" is calculated from the valve stem details. The stem diameter, pitch and lead are all needed for this calculation. These calculations will give the maximum torque needed by the valve, which can then be increased by a safety factor of between 20% to 100% depending on the age and duty of the valve. This torque is ultimately compared to the actuator manufacturer's torque tables to select an appropriately sized actuator.

Quarter-turn valves are harder to size. Some ball valve makers publish torque tables that allow users to enter a valve size and pressure class, which together with a differential pressure will give the valve torque at various positions of the valve's stroke.

However, most quarter-turn valve makers do not publish their torque data. *As a rough guideline see appendix 14.1 for a table of nominal torques compiled from averaged historical experience.*

The available power sources at a particular site will determine the selection of fluid or electric power.

The desired speed and frequency of operation will determine the power and degree of modulating capacity needed.

Depending on the force required for the valve, if a fail to position on loss of power mode is needed, the bias in actuator choice is toward fluid-powered actuators. There are some fail-to-position electric actuators, but these are limited to the small-size category.

Once the actuator type and size have been selected, then the other attributes can be defined, such as enclosure type and certification, if needed.

11.3.2   Retrofit actuator mounting hardware

In most retrofit applications a special mounting piece and stem coupling will need to be designed and fabricated. Although some specialist manufacturing companies produce mounting brackets and other hardware, this is usually the domain of a valve automation specialist.

Figure 11.3.2.a. Mounting hardware for a ball valve retrofit. The static spool piece is on the sling and the dynamic stem coupling is on the valve stem.
Image by permission of Rotork®.

There are many instances, such as old valve designs, where the details of the valve top-works have to be measured and photographed to allow custom-built hardware. The types of valve mounting hardware have been described in chapter 10. The majority of retrofits can be accommodated with the more common type of hardware. However, there are some applications that require more creative solutions.

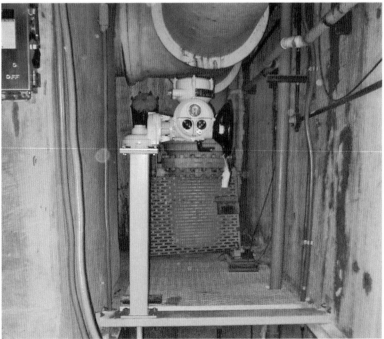

Figures 11.3.2.b.c.d. In this application an old water-powered actuator on a butterfly valve in a pit was replaced with a new electric actuator. Due to space restrictions, a compact quarter-turn gearbox was located on the valve stem and a drive shaft was connected to a multi-turn actuator at grade level to drive the gearbox.

Image by permission of Rotork®.

To facilitate the recording of valve top-works measurements, there are some tools that can be used. *See appendix 14.9.*

Chapter 12

Actuator maintenance

— ◆ —

One of the design objectives of valve actuators is to keep the maintenance needed to a minimum. A key requirement of an actuator is that it must be capable of remaining inactive in a hostile or remote environment for extended periods of time. Then, when it's required to move the valve, it must do so immediately and reliably.

The biggest enemy to reliability in actuators is corrosion. This can impact the sealing surfaces of a fluid power actuator or the electrical components and contacts of an electric actuator. In both cases the problem is compounded by inactivity.

For that and other reasons, critical safety shutdown valves are often tested by partially stroking the valves regularly to ensure they will respond to a control signal. This diminishes the effects of any accumulated static friction caused by local corrosion or media buildup in the valve.

Most facilities have a formal maintenance program designed to ensure that equipment is available for operation and is functioning correctly.

12.1 Fluid power actuators

The main concern with fluid power actuators is the integrity of the elastomeric seals on the pistons or internal moving elements. The life of a seal is dependent on several factors. For example, extremes of ambient temperature will impact how long they last. In particular, high temperatures will tend to vulcanize most common sealing materials causing them to harden, become brittle and be vulnerable to cracking.

The other concern for fluid power actuators is the integrity of the fluid power medium. For pneumatic actuators the quality of the instrument air impacts, not only the life of the sealing surfaces, but also the reliability of the more delicate controls and positioning devices. Moisture and particulates can corrode or block the small orifices needed to control the pneumatically automated valve in direction control valves and positioners.

Figure 12.1.a. Fluid power actuator maintenance.
Image by permission of Rotork®.

For these reasons, the maintenance required on the instrument air supply is a critical factor in ensuring the reliability of pneumatically automated valves.

Similarly, hydraulically automated valves, particularly those that use a central hydraulic system, are dependent on the purity and integrity of the hydraulic fluid.

Preventive maintenance for these valves would normally be focused on the central hydraulic system or the instrument air system, ensuring filters and dehydration equipment are fully functional and any leakage from the supply pipework is minimized. The actuators for these valves usually have a recommended life for elastomeric seals, particularly the dynamic seals on pistons and piston rods. For spring diaphragm actuators the critical component for maintenance is the diaphragm. In extreme applications these may need to be replaced every year or two, but in benign conditions they could last 5 to 10 years or more.

Pneumatic control valve assemblies with positioners require periodic recalibration. This is due to calibration drift from mechanical seat wear (zero point calibration) and changes in the physical properties of the internal mechanisms such as the flapper nozzle assembly. Some smart positioners allow calibration to be done remotely from the instrument shop via the integrated digital communications system.

12.2 Electric actuators

Light-duty smaller isolating actuators used in industrial applications are often located inside buildings and not exposed to the elements. Little or no maintenance is required or provided for these types of actuators and they are often inexpensive and easily replaced in case of failure.

Figure 12.2.a. Electric actuator maintenance.
Image by permission of Rotork®.

When specified correctly, the enclosure of a heavy-duty electric actuator will protect it from its working environment. The sealing technology employed on current actuators, even for hazardous area-certified units, precludes the intrusion of dirt and moisture.

In theory, little or no maintenance should be required for the first several years of an actuator's life. However, not every installation is performed correctly so sometimes an actuator may become contaminated with dirt or moisture. The other area of vulnerability is the electrical connections; moisture can find its way into the actuator via the electrical conduit if it's not properly sealed during installation. A post-installation examination should eliminate that problem, but a preventive maintenance check every two years is often recommended. This would include an inspection for physical damage, replacement of cover seals, a look at lubrication and inspection of full functional controls. A typical actuator examination would include the following:

a) *Inspect exterior for mechanical wear/damage. Check security of mounting bolts.*

b) *Engage hand drive and check function of hand operation.*

c) *Check lubricant level and condition in gear case, replenish as necessary.*

d) *Run actuator fully in both directions looking for correct operation of limit/torque switches and position indication.*

e) *Electrically isolate actuator. Remove all covers and check for moisture ingress or contamination.*

f) *Check condition of motor and contactor.*

g) *Check all internal electrical components for security and condition of plugs/sockets and cable assemblies.*

h) *In the case of specific explosive area actuators, check condition of all flame paths.*

i) *Replace all covers using new seals.*

j) *Grease the drive bush, where applicable. Check on condition of valve stem.*

k) *Remove electrical isolation and recheck operation in both local and remote operation.*

l) *List any parts that require replacement or show signs of wear.*

12.3 Predictive maintenance

Actuators that have digital communications capability, whether electric or fluid-powered, have some capacity to send diagnostic information to an asset management system. These systems are often capable of predicting when maintenance is required, not only on the actuator, but also on the valve itself.

For example, pneumatic smart positioners can measure the air pressure at various points within the positioner, such as the inlet from the supply and the outlet to the actuator, to ensure the actuator is available for operation. In addition, the air pressure in the actuator power chamber, (cylinder or diaphragm) can be used to determine the force or torque demand on the valve.

At the beginning of section 6.2, we discussed that this is a direct relationship in double-acting actuators; but with spring return actuators, the opposing force of the spring has to be factored into the calculation. To calculate the offset that the opposing spring force has on spring return actuators, the actuator position must be known so the associated compression of the spring at that positon can be determined. A position sensor measures position, usually on a continual basis.

On electric actuators the torque sensing mechanism measures the direct torque demand. A position sensor measures position, usually on a continual basis.

For both electric-powered and fluid-powered actuators, these two pieces of information, force and positon, allow maintenance personnel to monitor the condition of the valve.

Any increase or decrease in force demand can be seen, as well as the position at which the change occurs in the valve travel.

In addition, the total travel of the valve stem can be summed to give an indication of packing wear.

When a valve is installed and in operation, a base line "footprint" of the valve can be recorded. This footprint is a graph of force demand against position. As the valve ages, any changes in the force demand can be seen by comparing new footprints to the original.

The forces on a valve can be broken down into the following components:

- Valve sealing or packing friction
- Valve shaft-bearing friction
- Closure element seating friction
- Closure element travel friction
- Hydrodynamic force on closure element
- Valve stem piston effect
- Valve stem thread friction

Many of these elements are present in all types of valves, in varying degrees of magnitude. For example, the closure element travel friction in a butterfly valve is negligible, because it is not touching anything. Conversely, in a lubricated plug valve, this friction would be significant because the plug and body are in constant contact.

The analysis of the changes in these forces in a valve can give valuable information to plant maintenance planners on potential problems that could impact on the function of the valves. Seat refurbishment, packing replacement, stem lubrication and many other maintenance requirements can be predicted with reasonable accuracy by using the information collected by smart valve actuators.

Typical forces acting on a wedge gate valve
Valve seal or packing friction – function of stem diameter.
Valve shaft bearing friction – negligible.
Closure element on seat friction – wedge effect.
Closure element in-travel friction – negligible.
Hydrodynamic force on closure element in-travel – negligible.
Stem piston effect – function of pressure; negligible below 1.000 psi.
Valve stem thread friction – function of thread form and lubrication.

Figure 12.3.a. An example of some of the forces acting on a wedge gate valve.

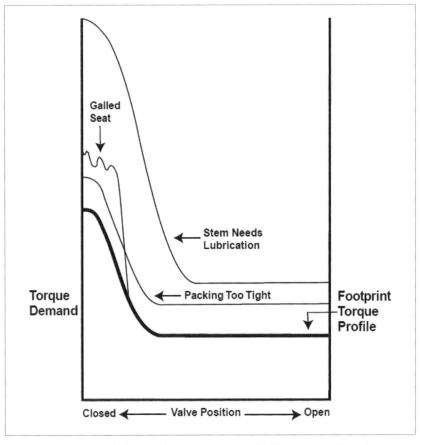

Figure 12.3.b. Valve problems can by identified by studying the valve torque demand in comparison to the reference profile.

In some instances separate asset management software can collect the data on the valves via digital communications systems and store it in a central database. Alternatively, some actuators can store their own data for download and analysis on spreadsheets or via product specific software.

Further information on predictive maintenance can be found on the resource page of www.cplloydconsulting.com.

Chapter 13

Applications

— ◆ —

13.1 Power plant

Figure 13.1.a. Power plant.
Image by permission of Rotork®.

There are many types of power plant, but a majority of the larger kind use a steam generating cycle. To operate effectively, the steam cycle requires a large number of automated valves both electric and fluid-powered.

Almost all non-renewable electricity for main power distribution is generated from a turbine/alternator combination. The turbine is often a steam-driven turbine, although gas turbines are also frequently used. Plants that use both gas and steam are known as "combined cycle" plants.

The turbine/alternator combination is the common factor that unites almost all power plants including some renewable energy types such as solar concentrator plants.

The difference between some of them is in the many methods of generating steam to run the steam turbines.

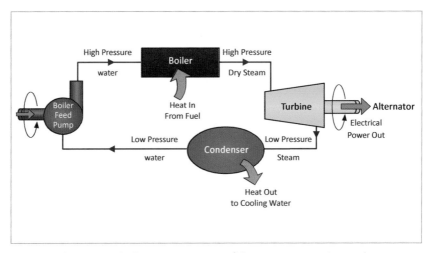

Figure 13.1.b. Basic components of the steam-generating cycle.

Steam is raised in boilers by applying a heat source to pressurized water. The water is pressurized by a boiler feed pump, which is essential for pushing the water through the boiler where it turns to steam. The boiler feed pump, by pushing the water, also pushes the steam through the turbine. There, the energy given it by the heat source is turned into mechanical energy by the turbine. This mechanical energy is then turned into electric energy by the alternator, which is driven by the turbine.

Figure 13.1.c. Electric actuators on steam lines in a power plant.
Image by permission of Rotork®.

The heat source could be the burning of bio-fuel, coal, oil or gas, or it could be nuclear fission or solar radiation.

Automated valves play a role in all of these processes.

First, the common element, the turbine and its coupled alternator (often called the turbo/alternator), can be looked at as a sub-process involving incoming steam from the boiler, expansion in the turbine and finally condensation of the steam in a condenser.

The turbine is invariably raised above the condenser on a substantial concrete structure (with the exception of saddle condensers, which are not common). This structure is often called the "turbine island." At various points in the expansion through the turbine, steam is bled off to provide preheating of the boiler feed water. Around the turbine are many small valves for this bled steam as well as drain valves and valves for other services such as lubrication and cooling. The alternator, that the turbine drives, is usually contained within a sealed enclosure filled with hydrogen to reduce wind resistance. This hydrogen has its own cooling system with more automated valves.

The condenser removes any residual heat from the steam after the last stage of the turbine. This turns the wet steam into water so it can be returned to the boiler by the boiler feed pump. A large amount of cold water is needed to cool the steam in the condenser. This water may come from a river or the sea or from cooling towers. Some of the largest valves, usually butterfly valves, are found at the inlet and outlet to the condenser. They are often electrically actuated.

Figure 13.1.d. Electric actuators on large water service butterfly valves.
Image by permission of Rotork®.

Turbo/alternators vary in size from a few megawatts (MW) to over 1,000 MW, and there could be several turbo/alternator sets in a power plant. The larger the size of the power plant, the more automated valves will be associated with the installation.

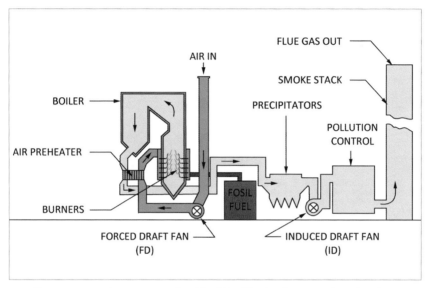

Figure 13.1.e. Air circuit in a fossil fuel-fired boiler showing FD and ID fans.

Conventional boilers mainly consist of a large hollow structure with water tubes lining the walls. Fuel is burned in the center of the structure to produce heat, which transfers to the water in the tubes. To produce rapid and efficient combustion, two fans are attached to the boiler: one pushes the air for combustion into the boiler (forced draft fan or FD) and the other sucks out the flue gas (induced draft fan or ID). The incoming air is usually taken from the top of the boiler house, where it is warmest. It may also pass through an air preheater that takes heat from the exhaust or flue gases and transfers it to the incoming air. The incoming air supports the fuel combustion in the middle of the boiler and passes over "superheater" tubes that further raise the temperature of the steam to make it "dry." The spent flue gas is then sucked out by the ID fan and passes to the stack for exhaust to the atmosphere.

It is important that the pressure in the boiler is maintained slightly below atmospheric pressure. To ensure the containment of the hot and toxic combustion gas. An excessive vacuum, however, could collapse the flue ductwork, so a careful balance between FD and ID fans is critical. This balance can be controlled by fan speed or by the positioning of the FD and ID fan dampers. These dampers are often controlled by electric or hydraulic actuators.

Figure 13.1.f. Electric actuator on a damper drive application.
Image courtesy of Harold Beck & Sons Inc.

Around the boiler are many automated valves controlling various processes such as the temperature of the superheated steam, boiler blowdown, make-up water and the fuel management valves.

Different fuels require different process systems. Gas and oil plants require valves and actuators to manage the flow of fuel to the burners located in the sides of the boiler. Heavy oil is sometimes used in power generation and may need heating and filtration before being pumped under pressure to the burners.

For new construction the preferred way of transforming gas fuel into electricity is to use a combined cycle plant. Compared to single-cycle, conventional plants, these have a smaller footprint, take less time to permit and construct, have lower emissions and have a higher thermal efficiency. A gas turbine is coupled to an alternator, and the exhaust from the turbine is ducted into a heat recovery steam generator (HRSG). This is a boiler designed specifically to recover the waste heat from the gas turbine. The HRSG raises steam, which is sent to a steam turbine and alternator combination. So there are two alternators associated with the combined cycle plant. The overall efficiency of these types of plant can be around 60% compared to a conventional plant at about 40%.

Coal plants often use pulverized fuel (PF), the coal is crushed to the consistency of talcum powder by ball mills. Fans and ducting are used to blow the PF up to the burners in the firewall of the boiler. This means that many more dampers and actuators are needed in a PF-fired power plant. At the back end of the boiler gas path, there are pollution controls such as electrostatic precipitators that collect the flue particulate. There are also plant sections that remove or neutralize the toxic elements such as sulfur in the flue gas. All these plant sections require automated valves and dampers.

Nuclear plants have a completely different heat source from conventional fossil-fired plants, even though the turbo/alternators may be similar. There are many different types of nuclear steam-raising designs such as gas-cooled reactors, boiling water reactors, heavy water reactors and other designs. The valves and actuators required to operate these facilities are heavily regulated. Both fluid power and electric actuators are used extensively inside and outside these reactors. The actuators have to be designed and qualified to withstand the extremes that would occur in an accidental loss of coolant if they are to be used inside the reactor containment area. On top of the extremes of temperature and steam impingement, the actuator seals and wiring must be resistant to radiation deterioration. Also, the whole actuator assembly may need to be capable of remaining intact and operational during and after a seismic event. For these reasons, these actuators are unique in design and construction and are more costly to produce and support.

The number of actuators needed for each type of power plant varies significantly. Relatively speaking, the largest number of actuators are used in a coal-fired plant with an actuator for every 2 MW of generating capacity. That means a typical 2,000 MW plant has about 1,000 valve actuators. The combined cycle plants use less: an actuator for roughly every 4 MW, which means a typical 750 MW combined cycle plant will use about 200 actuators.

13.2 Potable water treatment

Figure 13.2.a. Potable water treatment plant.
Image by permission of Rotork®.

Historically, one of the first infrastructure projects of any city, town or village has been a potable water supply. That means many new plants are being built in areas of expanding population. Also, in developed areas of the world, many existing water treatment plants are old and need modernizing. Valve actuators play an important role in maintaining the quality of our water supply by contributing to the efficient operation of potable water filtration plants.

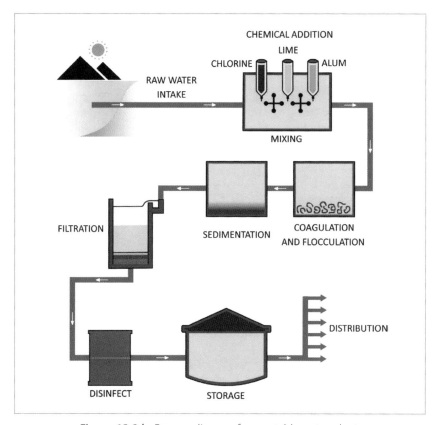

Figure 13.2.b. Process diagram for a potable water plant.

The process in a water plant takes raw water from rivers, lakes or other sources. A chemical coagulant is added to the water such as aluminum sulfate. The coagulant, when mixed into the water, causes the small solid particles to stick together in a process called flocculation. The water passes into large clarifiers that allow the solids in the water to settle to the base of the tank where they are collected for disposal.

The water then passes through an ozone or other type of disinfection process on its way to the filtration process.

At the filter bed, the flow of clarified water is introduced into the top of the filter along troughs. The water runs down through the filter media with large particles trapped first and finer particles adhering to the lower filter media. At the bottom of the filter, the bottom drain pipework collects the filtered water; it is passed on to the secondary disinfection stage, after which the water is stored ready for distribution.

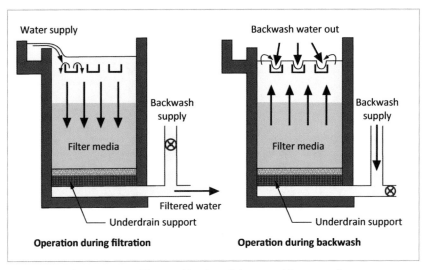

Figure 13.2.c. Filter and backwash in a potable water plant.

The process is continuous, with the exception of the filtration. These filters have to be cleaned regularly by reversing the flow to "backwash" the trapped particles out of the filter media to waste. This means the filter has to be taken out of the process (taken offline) while this is done. Water treatment plants have several filter beds so that while one is backwashed, the others can continue the process flow.

Automated valves are used in many locations around the plant to control the flow of water through the various processes using sluice gates, butterfly valves and gate valves with electric- or fluid-powered actuators. Usually the fluid-powered actuators are pneumatically powered, although decades ago many hydraulically powered actuators used the processed water under pressure.

Figure 13.2.d. Automated valves for backwash control.
Image courtesy of Auma.

A key automated valve is the rate-of-flow control valve. This valve is usually a butterfly valve and is modulated to ensure the rate of flow through the filter keeps the water level above the filter media but below the filter tank top. Sensors near the top of the filter sense high and low levels. The flow of water through the filter changes as solids build up and reduce the flow throughput. Because this would increase the filter water level, the rate-of-flow controller compensates by opening the rate-of-flow butterfly valve to increase effluent flow.

The most complex process is the backwash sequence needed to periodically remove solids from the filter. The backwash procedure drops the water level to just above the top of the media. The filter media is then agitated to loosen solids. This is done by pumping air up through the filter to loosen the filter media as well as the filtered solids, a process known as air wash or air scour. The top of the filter media may also be agitated by a "surface wash" system. Clean backwash water is then pumped up through the bottom of the filter by the backwash pumps. The solids are washed up and into the troughs at the top of the filter and out as waste. The inlet and outlet valves to the filter have to be sequenced carefully to ensure the correct flow; a sudden inrush of water could damage the filter by displacing the strata of the media. This sequence of events is usually done automatically by the process controller. Once the outflowing water reaches an acceptable level of clarity, measured by the turbidity sensor, the backwash is stopped and the valves are actuated back to their normal filtering position.

13.3 Waste water treatment plant

Figure 13.3.a. Waste water treatment plant.
Image by permission of Rotork®.

While safe drinking or potable water is a top priority, the other important piece of municipal infrastructure is the treatment facility for wastewater. Sanitary disposal of waste has contributed to the residents' health in all the cities of the world.

The process starts with the collection of waste using a sewer system. Often, automated valves and sluice gates will be used on the culverts and tunnels that connect the city buildings and houses with the wastewater treatment plant (WWTP).

On reaching the WWTP, usually by gravity flow, the wastewater is pumped up to a higher level by a lift or pump station. This allows the wastewater to progress through the plant via gravity flow. Bar screens are used at the lift station to catch trash such as diapers, rags and plastic bags.

The settling ponds or basins then allow solid waste to settle to the bottom where it is scraped away. The top layer is skimmed by "scum skimmers" to remove the scum and grease.

The water is then passed to the aeration basins where bacteria in the form of "activated sludge" is mixed with the wastewater. Air is blown into the mixture to accelerate the neutralization of the undesirable organic matter in the wastewater. The mixture of air and agitation speeds the reaction between the activated sludge and the organic waste matter.

Settling ponds then allow the organic clumps of sludge to settle to the bottom where some activated sludge is returned to the aeration basin to continue the reaction, while the rest is passed to the digester.

The digester takes the activated sludge and allows further reaction with the bacteria until it stabilizes; methane gas is a byproduct. Once the sludge has been stabilized, it can be disposed of or used for fertilization.

In some plants there may be a final filtration stage before the wastewater is returned to the environment via a river, lake or wetlands.

This filtration stage would be similar to the potable water filtration, but it's objective is to reduce to an acceptable level the biological oxygen demand (BOD) of the processed water. This BOD is a measure of the impact the discharged water will have on the environment.

The wastewater treatment plant uses a variety of automated valves including sluice gates on the channels and conduits, butterfly valves controling the air into the aeration basins and plug valves on the sludge pipelines. Valve actuators are also used on the tilting scum pipes at the primary settling ponds. The final filtration stage would also use actuators extensivly as these filters would need the automated valves to achieve level control as well as to facilitate the backwash procedure.

On large plants, several hundred valve actuators may be used. These actuators are not just on the basic process described above, but also on the many peripheral services needed such as chemical treatement for disinfection and residual clorine removal. Even the HVAC control of the administration and process control buildings use actuators.

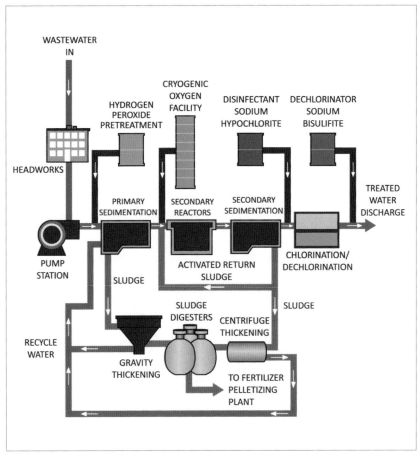

Figure 13.3.b. Process diagram for a wastewater treatment plant.

13.4 Oil and gas

13.4.1 Exploration and production

Figure 13.4.1.a. Offshore oil rig.
Image by permission of Rotork®.

Valve actuators of all types are used in the production of oil and gas, both on-shore and offshore. The basic processes for oil and gas production are similar regardless of the size of the field or its location.

The product coming up from the well contains a mixure of water, gas and oil in differing proportions. The first step in the process is to separate these elements. Vessels called separators are used to separate the oil, gas, and water components. The separated water will need to be further processed before it can be safely disposed of or reinjected into the well to enhance recovery.

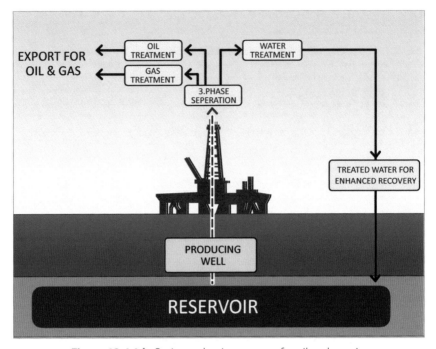

Figure 13.4.1.b. Basic production process for oil and gas rig.

Some contanimants may be included in the gas, such as hydrogen sulfide, carbon dioxide, helium or nitrogen; these need to be processed to bring the gas up to pipeline quality. Additionally, dehydration of the gas may be required. The oil may also need further stages of separation before it can be moved to the storage facility or export pipeline.

For well sites, the ancillary equipment required for production is extensive: not just the 3-phase separators (oil, water, gas), but also equipment for sulfur removable, glycol gas drying, gas compression, methanol injection, water re-injection, plus oil and gas metering.

On an offshore production platform, all this equipement has to be packaged into a small area. Also required are the main pumps and valves associated with pumping the product to onshore facilities or to floating storage vessels.

A process control system will control all of this equipment to ensure production optimization and the efficient running of the oil and gas production facility.

The number of valves associated with oil and gas production can be high depending on the size of the installation. Valves are needed to provide a variety of functions, from wellhead valves to separator level control to dehydration flow control and many other applications. Some valves perform critical safety shutdown functions, and therefore may require safety and integrity level (SIL) status along with their actuators.

In addition to this production equipment, there also may be accommodation facilities for personnel on an offshore platform. These require additional infrastructure, including auxiliary utility systems such as HVAC, power, water and fire safety, all with their own automated valves and dampers.

New oil wells are often self flowing. That means, the compressed gas cap on top of the oil in the natural reservoir provides sufficient pressure to force the product up the wellbore to the surface. However, whenever this pressure has been depleted, the enhanced recovery process may be required.

In this case, recovery is achieved by injecting water or gas into the well reservoir to force oil to the surface. The injection of the water or gas is carefully controlled to maximize production from the reservoir. This process requires a significant increase in the amount of equipment, including automated valves, required at and near the wellhead.

Oil and gas production valves	
Process Area	**Valve**
Wellhead	Chokes
	Upper/lower master
	Wing kill
	Injection
Export pipelines gas , oil	Isolation
Sectioning valves	Isolation
Separators primary, secondary	Level control
	Isolation
Gas scrubber and sulphur removal	Flow/level control
	Isolation
Gas dehydration	Flow/level control
	Isolation
Cooling water	Flow/level control
	Isolation

Process Area	Valve
Chemical injection	Flow/level control
	Isolation
Flare	Flow/level control
	Isolation
Utility and potable water system	Flow/level control
	Isolation
HVAC	Air dampers
	Water/steam control

Figure 13.4.1.c. Valves used on an oil production facility.

The technology used for hydraulic fracturing (hydrofracking) requires additional pipework to handle the introduction of the high-pressure hydraulic fluid into the well. Additional equipment is often needed to treat the returned fluid from the well before disposal or reuse. Because of this, more valves are used, some of which are automated.

13.4.2   Oil transportation

Figure 13.4.2.a. Oil pipeline.
Image by permission of Rotork®.

Once the oil has been processed from the wellhead to a suitable "pipline quality" product, it can be transported to the downstream refinery. This could be a journey of a few or thousands of miles. The transportation could take place through pipelines, by rail cars or by sea in super tankers and then eventually by road tanker.

All oil has to travel by pipline at some point, however. Oil is pushed through pipelines by pumping stations.

Oil pipelines and pump stations have many valves. Most terminals and pumping stations have storage tanks and a valve manifold system. This allows different products to be stored and transported through the pipeline system. Each tank will have its own valves, usually gate valves with electric actuators. The manifold system has many valves to allow the contents of the tanks to be routed to any of the pipelines. These manifolds are often automated with electrically operated double block and bleed valves.

The oil often passes through the possession of different owners and handlers. To make sure all parties know exactly how much oil has been transferred, it has to be measured by metering stations. This is part of the custody transfer process. The metering station often is constructed to be mounted on one or more skids for ease of transport and assembly at site.

A volumetric meter is at the center of the measurement procedure. These meter skids have large numbers of automated valves to connect to different pipelines. The meters have to be calibrated regularly, so "meter prover" loops are used for calibration. These are loops of pipe with a calibrated displacement are used to pass a known volume of oil through the meter to prove its accuracy. Special swiching valves are used for this procedure along with many other automated valves.

Once the oil is on its way through the pipeline, the surrounding environment has to be protected from accidental spills or ruptures in the pipeline. This is usually done by having shutdown valves to isolate the pipeline section if a leak is detected. These "sectioning" valves are often ball valves of the same bore as the pipeline. These valves can be very large with a high torque demand requiring very large shut-down actuators. In remote areas, powering these actuators can be a challenge. Electricity may not be available from a power grid, so compressed gas storage (nitrogen tanks) or solar power may have to be used.

High-pressure gas vane or scotch yoke actuators could be used with gas storage bottles. These need constant monitoring and regular replenishment of the gas bottles to ensure sufficient pressure and volume to close the valve when necessary.

Solar power can be used to drive a hydraulic motor coupled to a spring return hydraulic actuator. The spring provides the shutdown action to close the valve and the hydraulic motor opens the valve, but at a much slower speed based on the capacity of the solar power unit.

Environmental concerns have led to an increase in the use of these sectioning isolation valves to ensure any spills are quickly contained, particularly where pipelines cross rivers and streams.

13.4.3   Gas transportation

After the wellhead production, natural gas is processed through a separator to remove liquids. It may be further dehydrated using a glycol drier or by methanol injection. If there is sulphur in the gas, it will be removed by a desulphurization process. This is done because sulphur combined with water makes highly corrosive sulphuric acid. All these processes require valves, and they are often automated.

Once the gas is processed to pipeline quality, it is gathered from the wells at a compression station so it can travel down the main transportation pipeline. At the compression stations are either reciprocating or centrifugal compressors, often driven by the gas itself. These compressor stations often use quarter-turn actuators on plug or ball valves for their isolation and manifold valves. In addition, there are also blowdown valves that can vent the gas if necessary.

Figure 13.4.3.a. Valves on gas pipeline tapered plug valves.
Image courtesy of Emerson Bettis.

The isolating valve actuators are often the high-pressure, gas, hydraulic or gas-over-oil type of actuator, using the available pipline gas pressure for power.

The inlet and discharge valves to the compressor house are often low-pressure pneumatic or gas-driven actuators powered by a nominal 80 psi supply.

Main gas pipelines, just like oil pipelines, have to provide protection for the environment from leaks. Shutdown and isolating valves are used for this purpose.

In many areas of the world, the gas pressure in the pipline can be used to power these shutdown valve actuators. The energy to shut down the valve could be stored as gas pressure in a storage vessel. This gas would then be diverted into the actuator to drive the valve closed in an emergency.

Alternatively, energy could be stored as strain energy in the spring of a spring return actuator. Pipeline gas would be used to move the valve and actuator to the open position, while at the same time, compressing the actuator spring. In an emergency the gas pressure in the actuator would be released, allowing the spring to close the valve.

A break or leak in the gas pipeline would constitute such an emergency. This type of event is detected automatically by a line break detection system that monitors the rate of change of gas pressure in the pipeline. Slow changes in pipeline pressure are normal; but a fast change would usually indicate a line break and the shutdown process would be triggered.

In some countries the discharge of natural gas into the atmosphere is prohibited because it's a greenhouse gas. In those cases the type of actuator design that vents gas cannot be used. The alternative is to use an inert gas such as nitrogen as the stored energy. The drawback with this method is that it is dependant on maintaining the stored pressure. This scenario is vulnerable to leakage; hence the pressure needs to be carefully and constantly monitored.

Using a renewable energy source such as solar power is becoming the standard solution for remotely located shutdown valves. The solar energy is stored in batteries, and when needed, it powers a hydraulic pump. The hydraulic fluid under pressure is then used to set a hydraulic spring return actuator in the open position, ready for the spring to close the valve in an emergency. The solar power system can occasionally "top up" the hydraulic fluid pressure automatically, should there be any small amount of leakage in the system

Both plug and ball valves are used extensivly in gas transportation and storage. Predominately fluid-powered valve actuators have been used in the gas transportation industry, but there are also applications for electric actuators.

In major cities in the U.S., some distribution systems inside the "city gates" use DC powered electric actuators for section ESD (emergency shut down). Should a fire or other emergency occur, these valves would shut off the gas supply to a section such as a full city block.

13.4.4   Oil refining

Figure 13.4.4.a. Oil refinery.
Image by permission of Rotork®.

Before crude oil is suitable for any useful purpose, it has to be refined. The refining process is a separation of the oil with differing boiling points into different fractions. This is done in a distillation tower. The heated crude oil is introduced into the base of the tower. At various heights in the tower, the distillate is drawn off with the lightest (Naptha) at the top and heaviest (residual fuel oil) at the base.

All around the distillation tower are multiple valves and actuators providing control of the many service processes involved in the refining. A large number of these valves are modulating process control valves; often these are pneumatically operated sliding stem or quarter-turn valves. These valves provide level, temperature, pressure and other control functions.

Some oil refineries can have over 2,000 electric actuators located at various sections of the plant. In addition, a similar number of pneumatic actuators could be performing continuous process control on the many associated processes. An example of the processes in an oil refinery is shown below. Each block would have several automated valves.

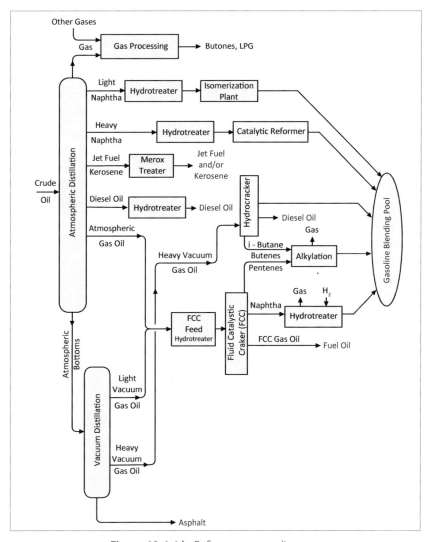

Figure 13.4.4.b. Refinery process diagram.

In addition to the distillation and associated peripheral processes involved in the refinery, there are pipelines and tankage for the incoming crude oil as well as for the multiple products produced by the refinery. There may also be custody transfer skids and even a dedicated power plant for the refinery. All in all, inumerable valves and actuators are associated with the oil refinery.

Chapter 14

Appendices

— ◆ —

14.1 Sizing

Valve sizing for gate and globe valves can be calculated using a combination of data from the valve, the media and some assumptions.

For a given valve size, if the stem details, media pressure and temperature are known, then the torque required to operate the valve can be estimated fairly accurately.

For a spreadsheet calculator, visit the website: www.cplloydconsulting.com. Look under the resources page.

Data needed	Units	Example
Valve bore size	**Inches**	6
Differential pressure	psi	100
Line pressure	psi	100
Stem diameter D	Inches	2
Stem lead	Inches	0.5
Pitch	Inches	0.5
Speed	Inches/minute	
Calculated data		
Area of valve	Square inches	28.27
Valve factor	From table	0.35
Mean stem dia		1.75
Pitch		0.5
Tan α	Lead/π *Dia	0.0909
Cos Θ	Cos thread angle	0.9681
Coeficient of friction	μ	0.14
Stem Factor		0.017
Thrust	Lbs	990
Piston effect	Lbs	314
Gland friction	Lbs	1000
Gland friction	Lbs	500
Total gland friction	Lbs	2500
Total thrust	Lbs	<u>3804</u>
Total torque	Lbs ft	<u>66</u>

$$\text{Stem factor} = D/24*(\text{Cos }\Theta*\text{Tan }\alpha+\mu)/(\text{Cos }\Theta-\text{Tan }\alpha*\mu)$$

Valve Factor				
Valve type	Gas below 1000F	Gas above 1000F	Liquid below 750F	Liquid above 750F
Solid wedge gate	0.45	0.5	0.35	0.4
Parallel slide	0.35	0.45	0.25	0.3
2" >Globe (screw down)	1.5	1.5	1.5	1.5
2" <Globe (screw down)	1.15	1.15	1.15	1.15

Figure 14.1.a. Example of sizing calculation for rising stem valve.

For quarter-turn valves the manufacturer's data is preferable. However there are some guidelines.

A nominal averaging of manufacturers of municipal butterfly valves for class 150 valves gives the following graphs.

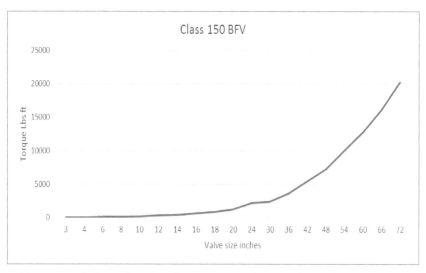

Figure 14.1.b. Butterfly valve, class 150 nominal torque demand sizes 3" to 72".

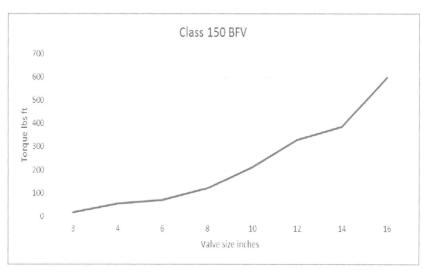

Figure 14.1.c. Butterfly valve, class 150 nominal torque demand sizes 3" to 16".

A rough guide for pipeline ball valves gives the following graphs:

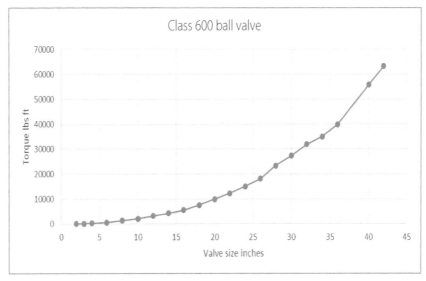

Figure 14.1.d. Pipeline ball valve, class 150 nominal torque demand sizes 2" to 42".

14.2 Valve stem buckling load

Where valves have a long, extended stem such as sluice gates, there is the potential for buckling the stem. This can be a problem for manual as well as motor-driven applications. A valve may be out of sight of the operator or just difficult to see. It may be difficult to determine when the valve is seated so extra force applied after seating could cause buckling of the stem. Once this occurs the valve is inoperable.

Figure 14.2.a. Motorized sluice gate with extended stem. Note the stem guides to reduce buckling risk.
Image by permission of Rotork®.

To reduce the risk of buckling, the valve stem may be supported at various places along its length.

The maximum allowable compressive force that can be applied is calculated using the formula below:

Maximum allowable force = $\pi 2EI/0.49L2$
(Euler formula assuming one end "fixed" in the gate and the other end "pinned" in the drive bush)

Buckling load		Example	
Diameter of stem	D	2	Inches
Radius of stem	r	1	Inches
Modulus of elasticity	I	30000000	psi for steel
Moment of inertia	πr4/4	0.7853975	
Unsupported length	L	60	Inches
Max force		3.17E+04	lbs. force

Figure 14.2.b. Sample buckling load calculation.

For a spreadsheet calculator, visit the website www.cplloydconsulting.com. Look under the resources page.

14.3 Certification bodies

There are many standards organizations that provide standards for electrical equipement that is to be located in hazardous and non-hazardous areas.

Actuators and accessories need to be certified by a third-party testing authority to demonstrate compliance to the appropriate standard for a specific environment and location. Frequently, local authorities determine the acceptable certification standards. For example, Australia will only accept actuators certified to the appropriate International Electric Commission (IEC) standard.

Name		Head office	Website
International Electric Commission	IEC	Switzerland	http://www.iec.ch/index.htm
European Committee for Electotechnical Standardization	CENELEC	Belgium	http://www.cenelec.eu/
Factory Mutual	FM	USA	http://www.fmglobal.com/default.aspx
Underwriters Laboratories	UL	USA	http://ul.com/
Canadian Standards Association	CSA	Canada	http://www.csagroup.org/
National Institute of Metrology, Quality and Technology	INMETRO	Brazil	http://www.inmetro.gov.br/english/
GOsudarstvennyy STandart	GOST	Russia	
National Electrical Manufacturers Association	NEMA	USA	http://www.nema.org
ATmosphères EXplosives	ATEX	EU	http://ec.europa.eu/growth/sectors/mechanical-engineering/atex/index_en.htm

Figure 14.3.a. Certifying bodies for hazardous area equipment.

14.4 Three-phase motors, direction of rotation and speed

The most common type of electric motor used in heavy-duty valve actuators is the 3-phase squirrel cage induction motor.

Each phase of an AC electrical supply can be described as a sinusoidaly varying voltage and current. In a single-phase supply, only one phase is present to carry power. To describe the phases in a 3-phase supply, it is common to use colors. In the U.S. the common colors used are black, red and blue with white as the optional neutral and green as ground or safety earth. The sequence in which the black, red and blue wires are connected to the motor are important because this dictates the direction of rotation of the magnetic field in the motor.

The typical valve actuator will change the direction of rotation of its motor by the use of a pair of reversing contactors. Typically, when one contactor is energized, the phase sequence will be A, B, C and when the other is energized, it will be C, B, A and the motor would spin in the opposite direction.

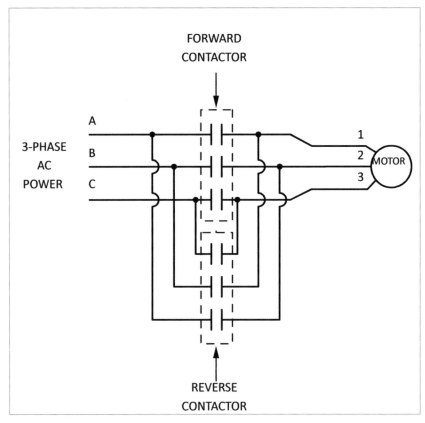

Figure 14.4.a. Schematic of reversing motor starter or contactor.

Valve actuators have position and torque monitoring devices that act to cut power to the motor, when they sense a position or torque limit. It is important to ensure the correct sequence of phase rotation of an actuator is followed during installation. Should the supply phases be sequenced incorrectly when connecting to the actuator terminals, then the actuator could run open when directed to run closed or visa versa. Compounding this problem is the fact the position and torque sensors used to cut power to the motor could be controlling or monitoring the wrong circuit. This can cause the full stall torque of the motor to be delivered to the valve, potentially causing severe damage.

To avoid this potential problem, many modern valve actuators are able to sense the 3-phase supply sequence and automatically select the correct contactor to energize based on the command signal. This automatic phase correction is often incorporated into the actuator firmware and is transparent to the installation and operations personnel.

The 3-phase motor speed is dependent on the number of pole pairs in the motor winding and the frequency of the electrical supply. For a motor with one pole pair connected to a 60 hertz (Hz) power supply, the speed of the rotating field in the windings will be 60 rotations every second or 3,600 revolutions per minute (rpm). Similarly, with a 50Hz supply, the speed will be 3,000 rpm.

If the motor winding is configured with two or more pole pairs, then the speed of rotation will be reduced by that factor. In other words, for a motor with two pole pairs operating with a 60Hz supply, the speed will be halved to 1,800 rpm.

The actual motor speed is slightly less than the speed of the rotating field and it's this slippage that induces an electro motive force (EMF) generated current in the rotor that drives the final torque output.

14.5 Hydrodynamic shock or "water hammer"

When a fluid is traveling through a pipeline it has its own momentum. The magnitude of this momentum depends on the fluid density and speed at which it is flowing. If this flow is rapidly arrested then the effect could be catastrophic, just like any object traveling at high speed that hits a solid object.

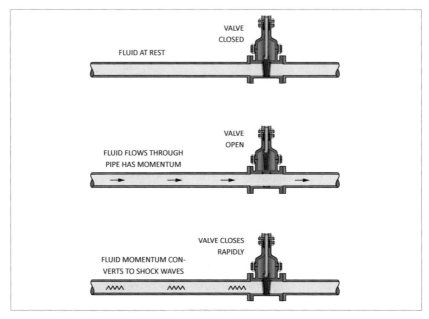

Figure 14.5.a. Water hammer in a pipeline.

If a valve is closed, too rapidly, there is the possibility that the energy of the fluid momentum could damage the valve and pipeline. This phenomenon is know as "water hammer" or hydrodynamic shock. The hammering effect is caused by shock waves traveling at the speed of sound in the fluid.

For applications with automated valves using high speed actuators, this possibility must be taken into consideration when selecting the actuator operating speed. In some applications a valve may need to open quickly; this in turn determines the power required of the actuator and the operating speed.

The speed of a fluid power actuator can be reduced in one direction by restricting the exhausting air flow or hydraulic fluid flow in the power cylinder for that direction.

Three-phase electric actuators, however, are usually fixed speed devices, so an alternative solution is to increase the closing stroke time of the valve by turning the electric actuator on and off. This is achieved by an integral timer in the actuator control that switches the motor starter on and off for selected intervals. This means that the valve can be closed over an extended period of time, allowing the fluid momentum in the pipeline to dissipate harmlessly.

There are formula that allow the pressure pulse to be calculated. For slow valve closure on a theroretical incompressible fluid:

F=ρLA (dv/dt)
Where in SI units:
F is the force generated by closure
ρ is the fluid density
L is the pipe length
A is the pipe cross-sectional area
v is the fluid velocity
t is the valve closure time.

For a spreadsheet calculator, visit the website: www.cplloydconsulting.com. Look under the resources page.

14.6 Galvanic insulation and cathodic protection

Corrosion on pipelines can cause leaks and spills of environmenally harmful liquids or gases, as well as disruption to infrastructure and loss of product and profits.

To counteract corrosion in steel pipelines, "cathodic protection" is used. This technology controls the corrosion of the pipeline metal by making it the cathode of a galvanic cell. A sacrificial anode is connected to the cathode. This anode then corrodes instead of the pipeline cathode. Sometimes an impressed current is used to enhance the sacrificial effect.

However, where electric actuators are used on a pipeline, they have to be connected to ground (or earthed) for safety purposes to comply with the electrical code. This means the impressed current of the cathodic protection system will dissipate through the ground connection.

To overcome this, the actuator may be insulated from the valve. This insulation must interpose between the valve and the actuator but still must transmit the drive force from the actuator to the valve stem.

It is easier to place this insulation between an actuator and a secondary gearbox because only torque and not thrust will need to be transmitted by the insulated drive. The thrust is then contained by the usual components of the secondary gearbox thrust base. Using a secondary gearbox also has the effect of reducing the tortional forces on the insulation.

A dielectric material such as nylon or teflon is used to insulate the static mounting surface and bolts between the actuator and secondary gearbox. A similar material is used on the rotating dynamic coupling. This effectively insulates the actuator from the valve and secondary gearbox and allows grounding of the electric actuator without impacting the cathodic protection or impressed current.

14.7 Air tank sizing

Pneumatic fluid power actuators may be required to move a valve to a certain position under failure conditions, even when there is the possibility of the loss of the instrument air supply.

Although a spring return actuator can perform this function, it may suit the appliction to use a double-acting actuator with an reserve air supply stored in a dedicated air accumulator.

The advantages of this configuration are that:

1) Control of the valve can be extended for more than one stroke. A spring return unit can only move to a fail fully open or a fail fully closed position and remain in that position until control and air power are restored.
2) A double-acting actuator is usually more compact and lighter than the equivalent spring return actuator of similar torque output. Generally, for the same size actuator body and cylinder diameter, a spring return configuration can only deliver about one-third of the torque of a similar double-acting actuator to move a valve.
3) The cost of a spring return actuator needed to operate a given valve size (especially larger valves) is often much more than the equivalently sized double-acting actuator.

To power the actuator on loss of instrument air, the double-acting actuator requires an air reservoir to store sufficient instrument air to move it through the required number of strokes. The air reservoir size has to be determined and this requires some calculation.

The information needed for this calculation is as follows:
1) The valve torque requirement at the end of stroke (assuming this is the maximum torque requirement). This information is supplied by the valve maker.
2) The site instrument air supply pressure.
3) The selected actuator torque output at the site instrument air supply pressure. This is found in the the actuator manufacturer's sizing chart.
4) The swept volume per stroke of the actuator. This is found in the actuator manufacturer's technical information.

With this information the adiabatic equation for gases can be used to calulate the required air receiver tank volume.

Where

Vr - is the air receiver tank volume

Vs - is the swept volume of the actuator per full stroke

Pi - is the instrument air pressure absolute (add 14.7 psi to normal gauge pressure).

Ps - is the pressure required by the actuator to deliver the required end of stroke torque. This is estimated from the ratio of valve demand torque divided by actuator catalogue torque at the instrument supply pressure multiplied by the instrument air supply absolute pressure.

These values are then substituted into the following equation:-

$$Vr = \frac{Vs}{\left[\dfrac{Pi}{Ps}\right]^{\frac{1}{nk}} - 1}$$

Where n is the number of strokes required from the actuator and k is the gas constant for instrument air (1.4).

An example of the calculation for a hypothetical 10" pipeline ball valve with a required closing torque demand of 24,000 inch-pounds (in-lbs) and an actuator with an output torque of 38,000 in-lbs at 8o psi and a swept volume of 850 cubic inches per stroke follows.

Calculation of air receiver size for pneumatic actuators			
Actuator swept volume	Vs	Cubic inches	850
Pressure needed at end of actuator stroke, absolute	Ps	psi Ab	65.23
Number of strokes required	n		2
Instrument air header pressure absolute	Pi	psi Ab	94.7
Gas constant for air	k	Ratio	1.4
Actuator sizing chart/inst air header pressure psi g	Pc	psi g	80
Max valve torque required at end of actuator stroke, from valve-maker data	Tv	In. lbs	24,000
Actuator end torque at 80 psi from actuator sizing data	Ta	In-lbs	38,000
Required volume of air receiver	VR	Cubic inches	5,968
Required volume of air receiver	VR	U.S. gallons	25.8

Figure 14.6.a. Air tank size calcuation.

For two strokes the required air receiver would have to be over 25.8 U.S. gallons in volume. A standard 30-gallon tank could be selected, or to increase the safety margin, a standard 60-gallon tank.

For a spreadsheet calculator, visit the website: www.cplloydconsulting.com. Look under the resources page.

14.8   Actuator specification and data sheets

To define an actuator, so that an automated valve supplier can provide the correct equipement to the purchaser of the automated valve assembly, several formats are used. The most common is a combination of a written specification and instrument data sheet.

The written specification is usually compiled by the consultant engineer based on that company's standard document modified by any specifics required by the end user or the specific requirements of that particular application. It describes the general requirements of the actuator for the project and is usually not related to the actuator size.

In contrast, the data sheet is specific to a particular valve tag number (unique identifier) or numbers and contains very specific information on the valve size, pressure, media and controls.

The relevant sections from a pneumatic actuator specification could look like this:

SPECIFICATION FOR PNEUMATIC OPERATORS
POWER MEDIUM: COMPRESSED AIR

1) GENERAL:

A) Cylinder design and material selection shall be compatible with the pneumatic driving medium, which shall be compressed air up to 150 psi.

B) Specification is intended to cover the design, construction, and fabrication of pneumatic scotch yoke-type cylinder operators for quarter-turn flow control devices.

C) For all double-acting actuators, the scotch yoke design is intended to give an end-of-travel torque of at least 1.5 times the mid-travel torque. That is, the torque at 0° travel will be 1.5 times the torque at 45° travel, and the torque at 90° travel (maximum travel) will be the same as the torque at 0°.

D) For all spring return actuators, the torque output of the pneumatic stroke shall have a similar magnitude and characteristic to the spring return stroke. On the pneumatic stroke, the break torque output at 0° shall be at least 1.5 times the end torque at 90°, and the torque at 45° shall be no less than 70% of the end torque. Similarly, on the spring return stroke, the break torque at 90° will be no less than 1.5 times the spring end torque, and the torque in the mid-position, at 45°, shall be no less than 70% of the spring end torque at 0°.

2) BODY DESIGN:

A) The valve actuator shall be designed to operate in hostile environments – outdoor and indoor installations. The center body shall be of a fully enclosed design to preclude the possibility of injury to personnel during operation.

B) The actuator shall be fitted with a visual position indicator easily understood and readable from a distance of approximately ten (10) yards (nine (9) meters).

C) Actuators shall have external, easily adjustable position stops. These stops are to be fully sealed to prevent leakage of lubricant from the center body.

D) For the mounting of accessories, tapped holes are to be provided on the front and rear face of the actuator, and on the body cover.

E) The center body cover is to be easily removable to allow for inspection of the center body without disassembling the entire unit or removing the unit from the valve.

3) SPRING CARTRIDGE:

A) Spring return units are to be of the spring cartridge type design. The spring cartridge will be a factory-sealed unit to prevent interference with the compressed spring onsite.

B) Spring cartridge design to be such that under no circumstances can the spring cartridge be removed from the actuator center body while in the compressed condition. Units must be depressurized and the spring in the contained preload condition before the cartridge can be dismounted.

4) CYLINDER CONSTRUCTION:

A) Pneumatic cylinder shall be of tie rod construction with a pressure rating of 150 psi minimum.

B) Cylinder barrels shall be carbon steel. The inside surface of the barrel shall have a 20-microinch finish.

C) Cylinder heads, caps, and pistons shall be ASTM pressure grade ductile iron.

E) Piston rod bushing shall be bronze.

F) Piston rod seal shall be elastomeric material suitable for air service.

G) Tie rod shall be of carbon steel.

5) CENTER BODY:

A) The center body shall be a one-piece ductile iron casting to maintain correct bearing alignment. A removable cover shall be provided to totally enclose the center body. This cover is to incorporate a watertight vent.

B) When a manual override is required, it should be designed to give the maximum torque output from the operator.

C) The manual override must be capable of being declutched to provide uninhibited power operation.

D) Actuators of lower torque output may be provided with a jackscrew-type manual override.

6) GENERAL REQUIREMENTS:

A) Actuator is to be capable of being mounted in any position.

B) When hydraulic override is provided, the customer is required to specify the valve stem orientation.

C) Unless otherwise specified, it will be assumed that the actuator is mounted horizontally and parallel to the pipeline.

D) Unless otherwise specified, the spring return actuators will be assembled spring close. That is, when the pneumatic cylinder is depressurized, the spring forces the output shaft to turn in the clockwise direction when viewed from above.

The relevant sections from an electric actuator specification could look like this:

SPECIFICATION FOR ELECTRIC OPERATORS

POWER MEDIUM: 3-PHASE ELECTRIC

A)   Each actuator shall include electric motor, reduction gearing, reversing starters, thermal overloads, controls transformer, limit controls, non-intrusive local controls and a digital 2-wire control system field unit as a complete integrated package to ensure proper coordination, compatibility, and operation of the system.

   1)   Provide actuators capable of setting of torque, turns, and configuration of indication contacts without the necessity to remove any electrical compartment covers.

   2)   Enclosure:

      a)   Watertight to IP68, BASEEFA classification. Enclosure must be certified NEMA 6 by FM (Factory Mutual), for all units except those in classified areas.

      b)   Provide explosion proof, NEMA 7 for actuators located in the following spaces:
      (1)   Wet wells
      (2)   Digester Complex

   3)   Provide an internal watertight compartment to protect switches, contacts, motor and internal electronics from ingress of moisture and dust when the external terminal cover is removed.

   4)   Provide each actuator with a handwheel for manual operation. Provide a hammer blow device, which permits the motor to come up to speed before picking up load and unseating valve.

B)   Motors:

   1)   Open/Close applications: motors, class F with 15-minute duty rating.

   2)   Modulating applications: motors, class H with a 30-minute duty rating.

   3)   Motor: low inertia, high-torque type to prevent over travel.

C) Provide internal clutch that cannot engage handwheel operating mechanism and motor-operating mechanism at the same time. Friction type declutching is not acceptable.

   1) Provide handwheel with arrow and the word CLOSE or SHUT cast on handwheel to indicate turning direction to close.

   2) Handwheel must not rotate during power operation.

   3) Provide handwheel and low gear ratio combined to give maximum rate of movement possible with 80 pounds (36 kilograms) rim pull.

D) Reduction Unit:

   1) Metal worm wheel and worm shaft-type.

   2) Provide an oil-filled gear box.

   3) Worm shaft to operate in ball or roller bearings and be machine cut, ground, and highly polished, hot-rolled steel, hardness 50-60 Rockwell Scale C bronze worm wheel with large contact area. Provide mating surfaces of dissimilar metals to prevent galling. Cast metals or gears manufactured from non-metallic materials are not acceptable.

   4) Worm and shafts to have heat-treated steel and be accurately machined. Output or driving shaft to operate in bronze bearing or in ball or roller bearings.

   5) Make provisions to take thrust in both directions.

   6) Gear caseto be cast iron or aluminum depending on size of actuator offered; all thrust or torque-bearing components shall be ductile iron.

   7) Provide drive bushing as part of a detachable thrust base making for easy retrofit.

E) Fully wire electric motor operators at factory and furnished complete with terminal strips for external power and control connections. Wiring should be copper with tropical grade PVC cover. Internal wiring to remain in a water tight compartment with external cover removed.

F) Provide manual or automatic control as indicated and specified.

G) Manual control should provide the following control, status, alarm and diagnostic capabilities locally, at the actuator:
   1) Control:
      a) Open/stop/close.
      b) Desired valve position control should be 0-100%.
   2) Status:
      a) Motor running open direction.
      b) Motor running close direction.
      c) Fully open.
      d) Fully closed.
      e) Percentage open 0-100% in 1% increments.
      f) Percentage output torque 0-100% in 1% increments.
   3) Alarms:
      a) Remote control communications failure.
      b) Actuator alarm.
      c) Valve alarm.

H) Automatic control: Provide remote automatic control as indicated.

I) Provide a backup power source integral to the actuator to ensure that in the event of a main power supply loss or failure, the indication contacts still function on change of status.

J) Provide contacts and operating parts made of non-corrodible metal and suitable for a sea atmosphere and for contact with H_2S.

K) Starters/transformers: Consists of two relay contactors, 3-pole, mechanically interlocked, reversing, with suitable arc suppressors.

   1. Provide electromechanical starter capable of OPEN/CLOSE sixty starts per hour. Size the solid state starter for modulating service at 1,200 starts per hour.
   2. Provide automatic phase correction.

Data sheets often describe the specifics of both valve and actuator:

CONTROL VALVE DATA SHEET

Project		Data Sheet	of
Unit		Spec	
P.O.		Tag	
Item		DWG	
Contract		Service	
*MFR. Serial			

#			Units	Max Flow	Norm Flow	Min Flow	Shut-Off	
1	Fluid					Crit Press PC		
2			Units	Max Flow	Norm Flow	Min Flow	Shut-Off	
3		Flow Rate					--	
4		Inlet Pressure						
5	Service Conditions	Outlet Pressure						
6		Inlet Temperature						
7		Spec Wt/Spec Grav/Mol Wt					--	
8		Viscosity/Spec Heats Ratio					--	
9		Vapor Pressure P_v					--	
10		*Required C_v					--	
11		*Travel	%				0	
12		Allowable/*Predicted SPL	dBA	/	/	/	--	
13								

#								
14	Line	Pipe Line Size &	In	55		*Type (linear, part turn)		
15		Schedule	Out	56		Manufacturer		
16		Pipe Line Insulation		57		Model		
17		*Type		58		Maximum torque/thrust		
18		*Size		59		Minumum torque/thrust		
19		ANSI Class		60		IEC modulation rate		
20		Max Press/Temp		61		Fail position - loss of power		
21		*Mfr & Model		62		Fail Position - loss of signal		
22	Valve Body/Bonnet	*Body/Bonnet Matl		63	Actuator	Power supply voltage		
23		*Linear Material/ID		64		Stroke speed		
24		End	In	65		Close action torque/thrust or Limit		
25		Connection	Out	66		Maximum power demand		
26		Flg Face Finish		67		Orientation		
27		End Ext/Matl		68		Handwheel Yes or No		
28		*Flow Direction		69		Control signal /Comunications		
29		*Type of Bonnet		70		Comunication adress		
30		Lub & Iso Valve		71		Feedback signal		
31		* Packing Material		72		Actuator body material		
32		*Packing Type		73		Actuator mounting bracket		
33	Trim	*Type		74		Conduit entry size		
34		*Size		75				
35		Rated Travel		76	Actuator switch			
36		*Characteristic		77		End of travel switches		
37		*Balanced/Unbalanced		78		ESD contacts		
38		*Rated		79		Availabilty status signal		
39		C_v F_L X_r		80				
40		*Plug/Ball/Disk Material		81				
41		*Seat Material		82				
42		*Cage/Guide Material		83				
43		*Stem Material		84				
44	Specials/Accessories			85				
45		NEC Class		86	Tests	*Hydro Pressure		
46		Group		87		ANSI/FCI Leakage Class		
47		Div		88				
48				Rev	Date		Orig	App
49						Revision		
50								
51								
52								
53								
54								

* Information supplied by manufacturer unless already specified.

14.9 Retrofit tools

When a retrofit has to be done, the first visit to the site gives the retrofit technician the opportunity to collect data on the valve to be automated. If possible, measurements should be taken of the valve top-works. This will allow the valve adaption hardware to be manufactured from the data or at least facilitate a reasonable estimate of the cost of manufacturing the adaption.

Shown below are sample blank data collection sheets for multi-turn and quarter-turn valves.

For more retrofit tools, visit the website: www.cplloydconsulting.com. Look under the resources page.

CPLLOYD CONSULTING
Valves and Automation

Retrofit valve data collection sheet

Round flange top-works, multi-turn

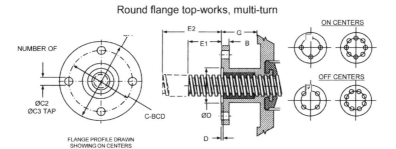

PLEASE FILL IN AS MUCH INFORMATION ON THE VALVE AS POSSIBLE

JOB No:		FLANGE DIAMETER	ØA	
SITE:		FLANGE THICKNESS	B	
MAKE / MODEL:		FLANGE No. HOLES	C1	
FIG / SER No:		FLANGE HOLE DIA	ØC2	
SIZE:		FLANGE HOLE THREAD	ØC3	
RATING:		BOLT CIRCLE	C-BCD	
TORQUE:		SPIGOT DIAMETER	ØD	
CYCLE TIME:		SPIGOT DEPTH	D	
No OF TURNS:		STEM HEIGHT VALVE CLOSED	E1	
MOV No:		STEM HEIGHT VALVE OPEN	E2	
TAG No:		VALVE TRAVEL	E3	
		PIPE FLANGE CLEARANCE	G	

COMMENTS:

MATCH THE STEM TYPE THAT BEST FITS YOUR APPLICATION AND FILL IN ALL RELEVANT INFORMATION

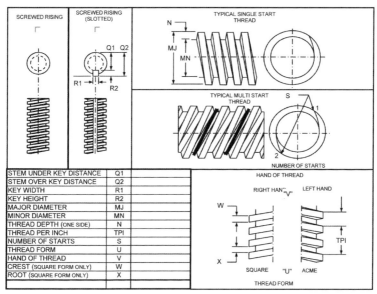

STEM UNDER KEY DISTANCE	Q1	
STEM OVER KEY DISTANCE	Q2	
KEY WIDTH	R1	
KEY HEIGHT	R2	
MAJOR DIAMETER	MJ	
MINOR DIAMETER	MN	
THREAD DEPTH (ONE SIDE)	N	
THREAD PER INCH	TPI	
NUMBER OF STARTS	S	
THREAD FORM	U	
HAND OF THREAD	V	
CREST (SQUARE FORM ONLY)	W	
ROOT (SQUARE FORM ONLY)	X	

CPLLOYD CONSULTING
Valves and Automation

Retrofit valve data collection sheet

Round flange top-works, quarter-turn

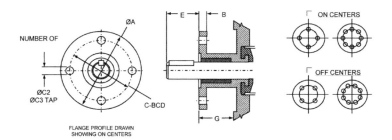

NUMBER OF

ØA

ØC2
ØC3 TAP

C-BCD

E B

ON CENTERS

OFF CENTERS

FLANGE PROFILE DRAWN
SHOWING ON CENTERS

G

PLEASE FILL IN AS MUCH INFORMATION ON THE VALVE AS POSSIBLE

Job No:		FLANGE DIAMETER:	ØA	
SITE:		FLANGE THICKNESS:	B	
MAKE / MODEL:		BOLT CIRCLE DIAMETER	C-BCD	
FIG / SER No:		FLANGE No OF HOLES:	C1	
SIZE:		FLANGE HOLE DIA:	ØC2	
RATING:		FLANGEHOLE THREAD:	ØC3	
TORQUE:		STEM HEIGHT FROM FLANGE:	E	
CYCLE TIME:		PIPE FLANGE CLEARANCE:	G	
No OF TURNS:				
MOV No:				
TAG No:				

COMMENTS:

MATCH THE STEM TYPE THAT BEST FITS YOUR APPLICATION AND FILL IN ALL RELEVANT INFORMATION

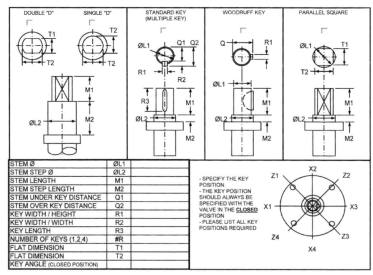

DOUBLE "D" SINGLE "D" STANDARD KEY (MULTIPLE KEY) WOODRUFF KEY PARALLEL SQUARE

STEM Ø	ØL1	
STEM STEP Ø	ØL2	
STEM LENGTH	M1	
STEM STEP LENGTH	M2	
STEM UNDER KEY DISTANCE	Q1	
STEM OVER KEY DISTANCE	Q2	
KEY WIDTH / HEIGHT	R1	
KEY WIDTH / WIDTH	R2	
KEY LENGTH	R3	
NUMBER OF KEYS (1,2,4)	#R	
FLAT DIMENSION	T1	
FLAT DIMENSION	T2	
KEY ANGLE (CLOSED POSITION)		

- SPECIFY THE KEY POSITION.
- THE KEY POSITION SHOULD ALWAYS BE SPECIFIED WITH THE VALVE IN THE **CLOSED** POSITION
- PLEASE LIST ALL KEY POSITIONS REQUIRED

X2
Z1   Z2
X1   X3
Z4   Z3
X4

45966837R00103

Made in the USA
Lexington, KY
17 October 2015